Green Chemistry

Green Chemistry

Edited by
Nate Newman

☰ Larsen & Keller
www.larsen-keller.com

Green Chemistry
Edited by Nate Newman
ISBN: 978-1-63549-139-5 (Hardback)

© 2017 Larsen & Keller

Larsen & Keller

Published by Larsen and Keller Education,
5 Penn Plaza,
19th Floor,
New York, NY 10001, USA

Cataloging-in-Publication Data

Green chemistry / edited by Nate Newman.
    p. cm.
Includes bibliographical references and index.
ISBN 978-1-63549-139-5
1. Green chemistry. 2. Chemistry. 3. Chemistry, Technical.
I. Newman, Nate.
TP155.2.E58 G74 2017
660--dc23

Table of Contents

Preface

This book is a valuable compilation of topics, ranging from the basic to the most complex theories and principles in the field of green chemistry. The design and development of procedures and technology that can reduce the emission of harmful gases into the environment is known as green chemistry. It seeks to implement efficient use of resources as well. Some of the diverse topics covered in this book address the varied branches that fall under this category. Green chemistry is an upcoming field of science that has undergone rapid development over the past few decades. This textbook will prove to be beneficial to graduates and post-graduates in the fields of environmental chemistry, renewable energy and ecosystem management.

A foreword of all Chapters of the book is provided below:

Chapter 1 - Green chemistry is also referred to as sustainable chemistry. It studies and focuses on the production of products that will minimize the use of hazardous substances. This chapter will provide an understanding of green chemistry; **Chapter 2 -** The metrics that are used to measure aspects of chemical processes are known as green chemistry metrics. Some of the other processes explained in the section are chemical process, natural-gas processing, galvanization, chemical garden, atom economy and supercritical hydrolysis. This chapter is an overview of the subject matter incorporating all the major aspects of green chemistry metrics and chemical processes; **Chapter 3 -** The problem solving approaches considered in green chemistry are alternatives assessment, advanced oxidation process, California green chemistry initiative and International Conference on Green Chemistry. Alternatives assessment is a problem solving approach that is used in environmental technologies and policies. The topics discussed in the chapter are of great importance to broaden the existing knowledge on green chemistry; **Chapter 4 -** The interdisciplinary aspects of green chemistry are chemical synthesis, process chemistry and biochemistry. Chemical synthesis is the inducement of a chemical reaction and this execution is done in order to obtain a product. In today's usage, this simply means that the process is reproducible and can be worked with in multiple laboratories. The major aspects of green chemistry are discussed in the following chapter; **Chapter 5 -** Condensation reaction is a chemical reaction and in this reaction, two molecules combine to form a large molecule. Other chemical reactions explained in the section are aldol condensation, Arndt-Eistert synthesis, Baeyer-Villiger oxidation, Dakin reaction, Simmons-Smith reaction, Heck reaction etc. The aspects elucidated in this section are of vital importance, and provides a better understanding of chemical reactions; **Chapter 6 -** Bioproducts are the materials and the chemicals that are obtained from renewable biological resources. The types of bioproducts that have been explained in the following section are bioenergy, biomass, biofuel, natural oil polyols, bio-based material etc. The topics discussed in the section are of great importance to broaden the existing knowledge on bioproducts; **Chapter 7 -** Environmental chemistry is the study of chemical phenomena that occur in the environment. Bioindicators, aquatic biomonitoring, environmental chemistry and pollinator decline are some of the topics that have been explained in the following chapter; **Chapter 8 -** Click chemistry is also known as tagging; it is not a particular reaction but it helps in describing a way to generate products that follow the examples of nature. Biocompatibility, bioconjugation and azide-alkyne huidgen cycloaddition are explained in the following section. The chapter focuses on all the aspects related to click chemistry and helps the readers in understanding the subject matter.

I would like to thank the entire editorial team who made sincere efforts for this book and my family who supported me in my efforts of working on this book. I take this opportunity to thank all those who have been a guiding force throughout my life.

Editor

Introduction to Green Chemistry

Green chemistry is also referred to as sustainable chemistry. It studies and focuses on the production of products that will minimize the use of hazardous substances. This chapter will provide an understanding of green chemistry.

Green chemistry, also called sustainable chemistry, is an area of chemistry and chemical engineering focused on the designing of products and processes that minimize the use and generation of hazardous substances. Whereas environmental chemistry focuses on the effects of polluting chemicals on nature, green chemistry focuses on technological approaches to preventing pollution and reducing consumption of nonrenewable resources.

Green chemistry overlaps with all subdisciplines of chemistry but with a particular focus on chemical synthesis, process chemistry, and chemical engineering, in industrial applications. To a lesser extent, the principles of green chemistry also affect laboratory practices. The overarching goals of green chemistry—namely, more resource-efficient and inherently safer design of molecules, materials, products, and processes—can be pursued in a wide range of contexts.

History

Green chemistry emerged from a variety of existing ideas and research efforts (such as atom economy and catalysis) in the period leading up to the 1990s, in the context of increasing attention to problems of chemical pollution and resource depletion. The development of green chemistry in Europe and the United States was linked to a shift in environmental problem-solving strategies: a movement from command and control regulation and mandated reduction of industrial emissions at the "end of the pipe," toward the active prevention of pollution through the innovative design of production technologies themselves. The set of concepts now recognized as green chemistry coalesced in the mid- to late-1990s, along with broader adoption of the term (which prevailed over competing terms such as "clean" and "sustainable" chemistry).

In the United States, the Environmental Protection Agency played a significant early role in fostering green chemistry through its pollution prevention programs, funding, and professional coordination. At the same time in the United Kingdom, researchers at the University of York contributed to the establishment of the Green Chemistry Network within the Royal Society of Chemistry, and the launch of the journal *Green Chemistry*.

Principles

In 1998, Paul Anastas (who then directed the Green Chemistry Program at the US EPA) and John C. Warner (then of Polaroid Corporation) published a set of principles to guide the practice of green chemistry. The twelve principles address a range of ways to reduce the environmental and health impacts of chemical production, and also indicate research priorities for the development of green chemistry technologies.

The principles cover such concepts as:

- the design of processes to maximize the amount of raw material that ends up in the product;
- the use of renewable material feedstocks and energy sources;
- the use of safe, environmentally benign substances, including solvents, whenever possible;
- the design of energy efficient processes;
- avoiding the production of waste, which is viewed as the ideal form of waste management.

The twelve principles of green chemistry are:

1. It is better to prevent waste than to treat or clean up waste after it is formed.

2. Synthetic methods should be designed to maximize the incorporation of all materials used in the process into the final product.

3. Wherever practicable, synthetic methodologies should be designed to use and generate substances that possess little or no toxicity to human health and the environment.

4. Chemical products should be designed to preserve efficacy of function while reducing toxicity.

5. The use of auxiliary substances (e.g. solvents, separation agents, etc.) should be made unnecessary wherever possible and innocuous when used.

6. Energy requirements should be recognized for their environmental and economic impacts and should be minimized. Synthetic methods should be conducted at ambient temperature and pressure.

7. A raw material or feedstock should be renewable rather than depleting wherever technically and economically practicable.

8. Reduce derivatives – Unnecessary derivatization (blocking group, protection/deprotection, temporary modification) should be avoided whenever possible.

9. Catalytic reagents (as selective as possible) are superior to stoichiometric reagents.

10. Chemical products should be designed so that at the end of their function they do not persist in the environment and break down into innocuous degradation products.

11. Analytical methodologies need to be further developed to allow for real-time, in-process monitoring and control prior to the formation of hazardous substances.

12. Substances and the form of a substance used in a chemical process should be chosen to minimize potential for chemical accidents, including releases, explosions, and fires.

Trends

Attempts are being made not only to quantify the *greenness* of a chemical process but also to factor in other variables such as chemical yield, the price of reaction components, safety in handling chemicals, hardware demands, energy profile and ease of product workup and purification. In one quantitative study, the reduction of nitrobenzene to aniline receives 64 points out of 100 marking it as an acceptable synthesis overall whereas a synthesis of an amide using HMDS is only described as adequate with a combined 32 points.

Green chemistry is increasingly seen as a powerful tool that researchers must use to evaluate the environmental impact of nanotechnology. As nanomaterials are developed, the environmental and human health impacts of both the products themselves and the processes to make them must be considered to ensure their long-term economic viability.

Examples

Green Solvents

Solvents are consumed in large quantities in many chemical syntheses as well as for cleaning and degreasing. Traditional solvents are often toxic or are chlorinated. Green solvents, on the other hand, are generally derived from renewable resources and biodegrade to innocuous, often naturally occurring product.

Synthetic Techniques

Novel or enhanced synthetic techniques can often provide improved environmental performance or enable better adherence to the principles of green chemistry. For example, the 2005 Nobel Prize for Chemistry was awarded, to Yves Chauvin, Robert H. Grubbs and Richard R. Schrock, for the development of the metathesis method in organic synthesis, with explicit reference to its contribution to green chemistry and "smarter production." A 2005 review identified three key developments in green chemistry in the field of organic synthesis: use of supercritical carbon dioxide as green solvent, aqueous hydrogen peroxide for clean oxidations and the use of hydrogen in asymmetric synthesis. Some further examples of applied green chemistry are supercritical water oxidation, on water reactions, and dry media reactions.

Bioengineering is also seen as a promising technique for achieving green chemistry goals. A number of important process chemicals can be synthesized in engineered organisms, such as shikimate, a Tamiflu precursor which is fermented by Roche in bacteria. Click chemistry is often cited as a style of chemical synthesis that is consistent with the goals of green chemistry. The concept of 'green pharmacy' has recently been articulated based on similar principles.

Carbon Dioxide as Blowing Agent

In 1996, Dow Chemical won the 1996 Greener Reaction Conditions award for their 100% carbon dioxide blowing agent for polystyrene foam production. Polystyrene foam is a common material used in packing and food transportation. Seven hundred million pounds are produced each year in the United States alone. Traditionally, CFC and other ozone-depleting chemicals were used in the production process of the foam sheets, presenting a serious environmental hazard. Flammable, explosive, and, in some cases toxic hydrocarbons have also been used as CFC replacements, but they present their own problems. Dow Chemical discovered that supercritical carbon dioxide works equally as well as a blowing agent, without the need for hazardous substances, allowing the polystyrene to be more easily recycled. The CO_2 used in the process is reused from other industries, so the net carbon released from the process is zero.

Hydrazine

Addressing principle #2 is the Peroxide Process for producing hydrazine without cogenerating salt. Hydrazine is traditionally produced by the Olin Raschig process from sodium hypochlorite (the active ingredient in many bleaches) and ammonia. The net reaction produces one equivalent of sodium chloride for every equivalent of the targeted product hydrazine:

$$NaOCl + 2\ NH_3 \rightarrow H_2N\text{-}NH_2 + NaCl + H_2O$$

In the greener Peroxide process hydrogen peroxide is employed as the oxidant, the side product being water. The net conversion follows:

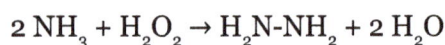

$$2\ NH_3 + H_2O_2 \rightarrow H_2N\text{-}NH_2 + 2\ H_2O$$

Addressing principle #4, this process does not require auxiliary extracting solvents. Methyl ethyl ketone is used as a carrier for the hydrazine, the intermediate ketazide phase separates from the reaction mixture, facilitating workup without the need of an extracting solvent.

1,3-Propanediol

Addressing principle #7 is a green route to 1,3-propanediol, which is traditionally generated from petrochemical precursors. It can be produced from renewable precursors via the bioseparation of 1,3-propanediol using a genetically modified strain of *E. coli*. This diol is used to make new polyesters for the manufacture of carpets.

Lactide

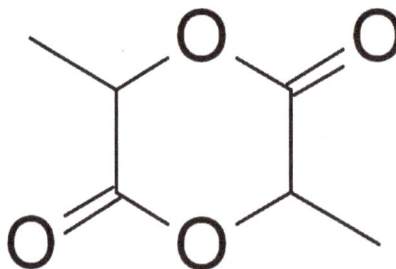

Lactide

In 2002, Cargill Dow (now NatureWorks) won the Greener Reaction Conditions Award for their improved method for polymerization of polylactic acid . Unfortunately, lactide-base polymers do not perform well and the project was discontinued by Dow soon after the award. Lactic acid is produced by fermenting corn and converted to lactide, the cyclic dimer ester of lactic acid using an efficient, tin-catalyzed cyclization. The L,L-lactide enantiomer is isolated by distillation and polymerized in the melt to make a crystallizable polymer, which has some applications including textiles and apparel, cutlery, and food packaging. Wal-Mart has announced that it is using/will use PLA for its produce packaging. The NatureWorks PLA process substitutes renewable materials for petroleum feedstocks, doesn't require the use of hazardous organic solvents typical in other PLA processes, and results in a high-quality polymer that is recyclable and compostable.

Carpet Tile Backings

In 2003 Shaw Industries selected a combination of polyolefin resins as the base polymer of choice for EcoWorx due to the low toxicity of its feedstocks, superior adhesion properties, dimensional stability, and its ability to be recycled. The EcoWorx compound also had to be designed to be compatible with nylon carpet fiber. Although EcoWorx may be recovered from any fiber type, nylon-6 provides a significant advantage. Polyolefins are compatible with known nylon-6 depolymerization methods. PVC interferes with those processes. Nylon-6 chemistry is well-known and not addressed in first-generation production. From its inception, EcoWorx met all of the design criteria necessary to satisfy the needs of the marketplace from a performance, health, and environmental standpoint. Research indicated that separation of the fiber and backing through elutriation, grinding, and air separation proved to be the best way to recover the face and backing components, but an infrastructure for returning postconsumer EcoWorx to the elutriation process was necessary. Research also indicated that the postconsumer carpet tile had a positive economic value at the end of its useful life. EcoWorx is recognized by MBDC as a certified cradle-to-cradle design.

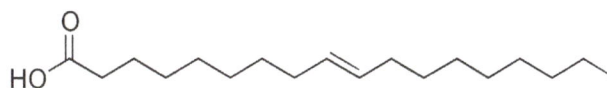

trans-Oleic acid

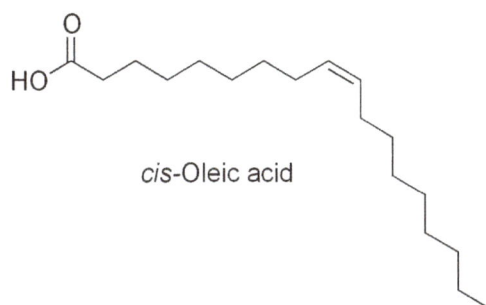

cis-Oleic acid

Trans and *cis* fatty acids

Transesterification of Fats

In 2005, Archer Daniels Midland (ADM) and Novozymes won the Greener Synthetic Pathways Award for their enzyme interesterification process. In response to the U.S. Food and Drug Administration (FDA) mandated labeling of *trans*-fats on nutritional information by January 1, 2006, Novozymes and ADM worked together to develop a clean, enzymatic process for the interesterification of oils and fats by interchanging saturated and unsaturated fatty acids. The result is commercially viable products without *trans*-fats. In addition to the human health benefits of eliminating *trans*-fats, the process has reduced the use of toxic chemicals and water, prevents vast amounts of byproducts, and reduces the amount of fats and oils wasted.

Bio-succinic Acid

In 2011, the Outstanding Green Chemistry Accomplishments by a Small Business Award went to BioAmber Inc. for integrated production and downstream applications of bio-based succinic acid. Succinic acid is a platform chemical that is an important starting material in the formulations of everyday products. Traditionally, succinic acid is produced from petroleum-based feedstocks. BioAmber has developed process and technology that produces succinic acid from the fermentation of renewable feedstocks at a lower cost and lower energy expenditure than the petroleum equivalent while sequestering CO_2 rather than emitting it.

Laboratory Chemicals

Several laboratory chemicals are controversial from the perspective of Green chemistry. The Massachusetts Institute of Technology has created the to help identify alternatives. Ethidium bromide, xylene, mercury, and formaldehyde have been identified as "worst offenders" which have alternatives. Solvents in particular make a large contribution to the environmental impact of chemical manufacturing and there is a growing focus on introducing Greener solvents into the earliest stage of development of these processes: laboratory-scale reaction and purification methods. In the Pharmaceutical Industry, both GSK and Pfizer have published Solvent Selection Guides for their Drug Discovery chemists.

Legislation

The EU

In 2007, The EU put into place the Registration, Evaluation, Authorisation, and Restriction of Chemicals (REACH) program, which requires companies to provide data showing that their products are safe. This regulation (1907/2006) ensures not only the assessment of the chemicals' hazards as well as risks during their uses but also includes measures for banning or restricting/ authorising uses of specific substances. ECHA, the EU Chemicals Agency in Helsinki, is implementing the regulation whereas the enforcement lies with the EU member states.

United States

The U.S. law that governs the majority of industrial chemicals (excluding pesticides, foods, and pharmaceuticals) is the Toxic Substances Control Act (TSCA) of 1976. Examining the role of regulatory programs in shaping the development of green chemistry in the United States, analysts have revealed structural flaws and long-standing weaknesses in TSCA; for example, a 2006 report to the California Legislature concludes that TSCA has produced a domestic chemicals market that discounts the hazardous properties of chemicals relative to their function, price, and performance. Scholars have argued that such market conditions represent a key barrier to the scientific, technical, and commercial success of green chemistry in the U.S., and fundamental policy changes are needed to correct these weaknesses.

Passed in 1990, the Pollution Prevention Act helped foster new approaches for dealing with pollution by preventing environmental problems before they happen.

In 2008, the State of California approved two laws aiming to encourage green chemistry, launching the California Green Chemistry Initiative. One of these statutes required California's Department of Toxic Substances Control (DTSC) to develop new regulations to prioritize "chemicals of concern" and promote the substitution of hazardous chemicals with safer alternatives. The resulting regulations took effect in 2013, initiating DTSC's *Safer Consumer Products Program.*

Green Chemistry Education

Many institutions offer courses and degrees on Green Chemistry. Examples from across the globe are Denmark's Technical University, and several in the US, e.g. at the Universities of Massachusetts-Boston, Michigan, and Oregon. A masters level course in Green Technology, has been introduced by the Institute of Chemical Technology, India. In the UK at the University of York University of Leicester, Department of Chemistry and MRes in Green Chemistry at Imperial College London. In Spain different universities like the Universidad de Jaume I or the Universidad de Navarra, offer Green Chemistry master courses. There are also websites focusing on green chemistry, such as the Michigan Green Chemistry Clearinghouse at www.migreenchemistry.org. Apart from its Green Chemistry Master courses the Zurich University of Applied Sciences ZHAW presents an exposition and web page "Making chemistry green" for a broader public, illustrating the 12 principles.

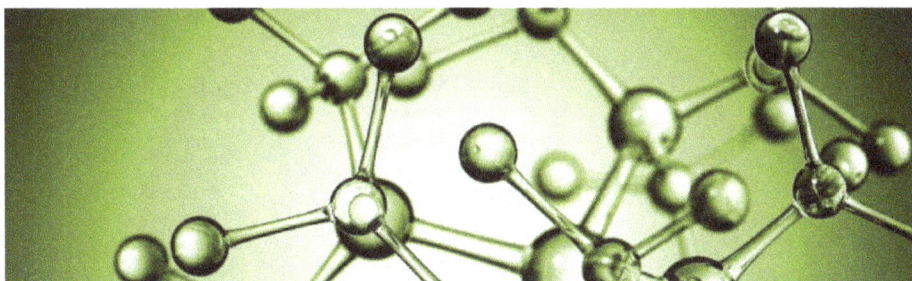

Contested Definition

There are ambiguities in the definition of green chemistry, and in how it is understood among broader science, policy, and business communities. Even within chemistry, researchers have used the term "green chemistry" to describe a range of work independently of the framework put forward by Anastas and Warner (i.e., the 12 principles). While not all uses of the term are legitimate, many are, and the authoritative status of any single definition is uncertain. More broadly, the idea of green chemistry can easily be linked (or confused) with related concepts like green engineering, environmental design, or sustainability in general. The complexity and multifaceted nature of green chemistry makes it difficult to devise clear and simple metrics. As a result, "what is green" is often open to debate.

Green Chemistry Awards

Several scientific societies have created awards to encourage research in green chemistry.

- Australia's Green Chemistry Challenge Awards overseen by The Royal Australian Chemical Institute (RACI).

- The Canadian Green Chemistry Medal.

- In Italy, Green Chemistry activities center around an inter-university consortium known as INCA.

- In Japan, The Green & Sustainable Chemistry Network oversees the GSC awards program.

- In the United Kingdom, the Green Chemical Technology Awards are given by Crystal Faraday.

- In the US, the Presidential Green Chemistry Challenge Awards recognize individuals and businesses.

References

- Sheldon, R. A.; Arends, I. W. C. E.; Hanefeld, U. (2007). "Green Chemistry and Catalysis". doi:10.1002/9783527611003. ISBN 9783527611003.

- Anastas, Paul T.; Warner, John C. (1998). Green chemistry: theory and practice. Oxford [England]; New York: Oxford University Press. ISBN 9780198502340.

- Cernansky, R. (2015). "Chemistry: Green refill". Nature. 519 (7543): 379. doi:10.1038/nj7543-379a.

- California Department of Toxic Substances Control. "What is the Safer Consumer Products (SCP) Program?". Retrieved 5 September 2015.

- Baron, M. (2012). "Towards a Greener Pharmacy by More Eco Design". Waste and Biomass Valorization. 3 (4): 395. doi:10.1007/s12649-012-9146-2.

Green Chemistry Metrics and Chemical Processes

The metrics that are used to measure aspects of chemical processes are known as green chemistry metrics. Some of the other processes explained in the section are chemical process, natural-gas processing, galvanization, chemical garden, atom economy and supercritical hydrolysis. This chapter is an overview of the subject matter incorporating all the major aspects of green chemistry metrics and chemical processes.

Green Chemistry Metrics

Green chemistry metrics are metrics that measure aspects of a chemical process relating to the principles of green chemistry. These metrics serve to quantify the efficiency or environmental performance of chemical processes, and allow changes in performance to be measured. The motivation for using metrics is the expectation that quantifying technical and environmental improvements can make the benefits of new technologies more tangible, perceptible, or understandable. This, in turn, is likely to aid the communication of research and potentially facilitate the wider adoption of green chemistry technologies in industry.

For a non-chemist the most attractive method of quoting the improvement might be *a decrease of X unit cost per kilogram of compound Y*. This, however, would be an oversimplification—for example, it would not allow a chemist to visualise the improvement made or to understand changes in material toxicity and process hazards. For yield improvements and selectivity increases, simple percentages are suitable, but this simplistic approach may not always be appropriate. For example, when a highly pyrophoric reagent is replaced by a benign one, a numerical value is difficult to assign but the improvement is obvious, if all other factors are similar.

Numerous metrics have been formulated over time and their suitability discussed at great length. A general problem observed is that the more accurate and universally applicable the metric devised, the more complex and unemployable it becomes. A good metric must be clearly defined, simple, measurable, objective rather than subjective and must ultimately drive the desired behavior.

Effective Mass Yield

Effective mass yield is defined as the percentage of the mass of the desired product relative to the mass of all non-benign materials used in its synthesis. Hudlicky *et al.* suggests the following equation:

Effective mass yield (%) = mass of products × 100 / mass of non-benign reagents

This metric requires further definition of a benign substance. Hudlicky defines it as "those by-products, reagents or solvents that have no environmental risk associated with them, for example, water, low-concentration saline, dilute ethanol, autoclaved cell mass, etc.". This definition leaves the metric open to criticism, as nothing is non-benign (which is a subjective term) and the substances listed in the definition have some environmental impact associated with them. The formula also fails to address the level of toxicity associated with a process. Until all toxicology data is available for all chemicals and a term dealing with these levels of "non-benign" reagents is written into the formula the effective mass yield is not the best metric for chemistry.

Carbon Efficiency

Carbon efficiency is a simplified formula developed at GlaxoSmithKline (GSK).

Carbon efficiency (%) = amount of carbon in product × 100 / total carbon present in reactants

This metric is a good simplification for use in the pharmaceutical industry as it takes into account the stoichiometry of reactants and products. Furthermore, this metric is of interest to the pharmaceutical industry where development of carbon skeletons is key to their work.

Atom Economy

Atom economy was designed in a different way to all the other metrics; most of these were designed to measure the improvement that had been made. Barry Trost conversely, designed atom economy as a method by which organic chemists would pursue "greener" chemistry. The simple definition of atom economy is a calculation of how much of the reactants remain in the final product.

For a generic multi-stage reaction:

1. $A + B \rightarrow C$
2. $C + D \rightarrow E$
3. $E + F \rightarrow G$

Atom economy = m.w. of G × 100 / Σ (m.w. A,B,D,F)

The drawback of atom economy is that assumptions have to be made. For example, inorganic reagents (such as potassium carbonate in a Williamson ether synthesis) are ignored as they are not incorporated into the final product. Also, solvents are ignored, as is the stoichiometry of the reagents.

The atom economy calculation is a very simple representation of the "green-ness" of a reaction as it can be carried out without the need for experimental results. However, it is useful as a low atom economy at the design stage of a reaction prior to entering the laboratory can drive a cleaner synthetic strategy to be formulated.

Reaction Mass Efficiency

Again developed by GSK, the reaction mass efficiency takes into account atom economy, chemical yield and stoichiometry. The formula can take one of the two forms shown below:

From a generic reaction where A + B → C Reaction mass efficiency = molecular weight of product C × yield / m.w. A + (m.w. B × molar ratio B/A)

Or more simply Reaction mass efficiency = mass of product C × 100 / mass of A + mass of B

Like carbon efficiency, this measure shows the "clean-ness" of a reaction but not of a process, for example, neither metric takes into account waste produced. For example, these metrics could present a rearrangement as "very green" but they would fail to address any solvent, work-up and energy issues arising.

Environmental (E) Factor

The first general metric for green chemistry remains one of the most flexible and popular ones. Roger A. Sheldon's **E-factor** can be made as complex and thorough or as simple as required. Assumptions on solvent and other factors can be made or a total analysis can be performed.

The E-factor calculation is defined by the ratio of the mass of waste per mass of product:

E-factor = total waste / product

As examples, Sheldon included E-factor analysis of various industries:

Table 1. E-Factors across the chemical industry			
Industry sector	**Annual production (t)**	**E-factor**	**Waste produced (t)**
Oil refining	10^6-10^8	Ca. 0.1	10^5-10^7
Bulk chemicals	10^4-10^6	<1–5	10^4-5×10^6
Fine chemicals	10^2–10^4	5–50	5×10^2–5×10^5
Pharmaceuticals	10–10^3	25–100	2.5×10^2–10^5

It highlights the waste produced in the process as opposed to the reaction, thus helping those who try to fulfil one of the twelve principles of green chemistry to avoid waste production. E-factors ignore recyclable factors such as recycled solvents and re-used catalysts, which obviously increases the accuracy but ignores the energy involved in the recovery (these are often included theoretically by assuming 90% solvent recovery). The main difficulty with E-factors is the need to define system boundaries, for example, which stages of the production or product life-cycle to considerm before calculations can be made.

Crucially, this metric is simple to apply industrially, as a production facility can measure how much material enters the site and how much leaves as product and waste, thereby directly giving an accurate global E-factor for the site. Table 1 shows that oil companies produce a lot less waste than pharmaceuticals as a percentage of material processed. This reflects the fact that the profit margins in the oil industry require them to minimise waste and find uses for products which would normally be discarded as waste. By contrast the pharmaceutical sector is more focussed on molecule manufacture and quality. The (currently) high profit margins within the sector mean that there is less concern about the comparatively large amounts of waste that are produced (especially considering the volumes used) although it has to be noted that, despite the percentage waste and E-factor being high, the pharmaceutical section produces much lower tonnage of waste than any other sector. This table encouraged a number of large pharmaceutical companies to commence "green" chemistry programs.

By incorporating yield, stoichiometry and solvent usage the E-factor is an excellent metric. Crucially, E-factors can be combined to assess multi-step reactions step by step or in one calculation.

Comparison of Metrics

A group of scientists at GSK attempted to compare the metrics currently available, with the goal of designing a metric set suitable for their business. Following lengthy comparisons of different metrics for twenty eight reaction types and comparing simple and complex costings for four different drug molecules, they came up with a number of conclusions. It was suggested that yield is still a very good metric especially for high value added chemistries such as those used in pharmaceutical synthesis, but conceded that yield does not encourage sustainable practices. Atom economy has its uses in conjunction with other metrics but is not suitable as a standalone green chemistry metric. Reaction mass efficiency combines both process and chemistry features and thereby has potential to be used optimally by chemists, process chemists and chemical engineers. The reaction mass efficiency is more likely to encourage sustainable practices as it focuses attention away from the waste and towards the use of materials. However, in terms of measuring the process and ease of use E-factors are the most effective calculation at the current time.

The EcoScale

The EcoScale is a recently developed metric tool for evaluation of the effectiveness of a synthetic reaction. It is characterized by simplicity and general applicability. Like the yield-based scale, the EcoScale gives a score from 0 to 100, but also takes into account cost, safety, technical set-up, energy and purification aspects. It is obtained by assigning a value of 100 to an ideal reaction defined as "Compound A (substrate) undergoes a reaction with (or in the presence of)inexpensive compound(s) B to give the desired compound C in 100% yield at room temperature with a minimal risk for the operator and a minimal impact on the environment", and then subtracting penalty points for non-ideal conditions. These penalty points take into account both the advantages and disadvantages of specific reagents, set-ups and technologies. By calculating the EcoScale, a quick assessment of the "greenness" of reaction protocols is obtained, and the areas that need further attention are clearly indicated, which finally can drive improvement of reaction conditions.

Chemical Process

In a scientific sense, a chemical process is a method or means of somehow changing one or more chemicals or chemical compounds. Such a chemical process can occur by itself or be caused by an outside force, and involves a chemical reaction of some sort. In an "engineering" sense, a chemical process is a method intended to be used in manufacturing or on an industrial scale to change the composition of chemical(s) or material(s), usually using technology similar or related to that used in chemical plants or the chemical industry.

Neither of these definitions is exact in the sense that one can always tell definitively what is a chemical process and what is not; they are practical definitions. There is also significant overlap in these two definition variations. Because of the inexactness of the definition, chemists and other scientists use the term "chemical process" only in a general sense or in the engineering sense. However, in the "process (engineering)" sense, the term "chemical process" is used extensively.

Although this type of chemical process may sometimes involve only one step, often multiple steps, referred to as unit operations, are involved. In a plant, each of the unit operations commonly occur in individual vessels or sections of the plant called units. Often, one or more chemical reactions are involved, but other ways of changing chemical (or material) composition may be used, such as mixing or separation processes. The process steps may be sequential in time or sequential in space along a stream of flowing or moving material. For a given amount of a feed (input) material or product (output) material, an expected amount of material can be determined at key steps in the process from empirical data and material balance calculations. These amounts can be scaled up or down to suit the desired capacity or operation of a particular chemical plant built for such a process. More than one chemical plant may use the same chemical process, each plant perhaps at differently scaled capacities. Chemical processes like distillation and crystallization go back to alchemy in Alexandria, Egypt.

Such chemical processes can be illustrated generally as block flow diagrams or in more detail as process flow diagrams. Block flow diagrams show the units as blocks and the streams flowing between them as connecting lines with arrowheads to show direction of flow.

In addition to chemical plants for producing chemicals, chemical processes with similar technology and equipment are also used in oil refining and other refineries, natural gas processing, polymer and pharmaceutical manufacturing, food processing, and water and wastewater treatment.

Unit Processing in Chemical Process

Unit processing is the basic processing in chemical engineering. Together with unit operations it forms the main principle of the varied chemical industries. Each genre of unit processing follows the same chemical law much as each genre of unit operations follows the same physical law.

Chemical engineering unit processing consists of the following important processes:

- Oxidation
- Reduction
- Hydrogenation
- Dehydrogenation
- Hydrolysis
- Hydration

- Dehydration
- Halogenation
- Nitrification
- Sulfonation
- Ammoniation
- Alkaline fusion
- Alkylation
- Dealkylation
- Esterification
- Polymerization
- Polycondensation
- Catalysis

Academic Research Institutes in Process Chemistry

Institute of Process Research & Development, University of Leeds

Electrolysis

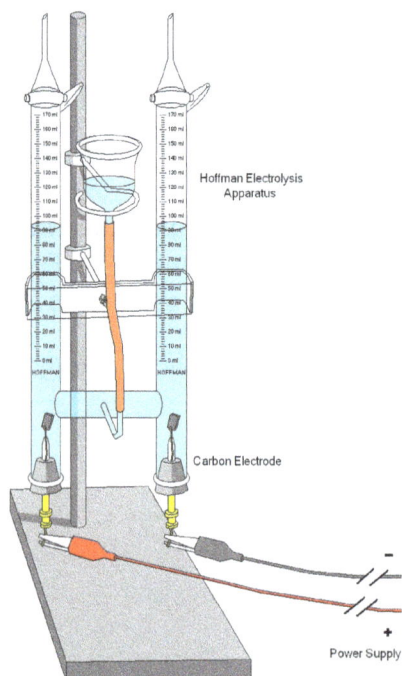

Illustration of an electrolysis apparatus used in a school laboratory.

In chemistry and manufacturing, electrolysis is a technique that uses a direct electric current (DC) to drive an otherwise non-spontaneous chemical reaction. Electrolysis is commercially important as a stage in the separation of elements from naturally occurring sources such as ores using an electrolytic cell. The voltage that is needed for electrolysis to occur is called the decomposition potential.

History

The word "electrolysis" was introduced by Michael Faraday in the 19th century, on the suggestion of the Rev. William Whewell which since the 17th century was associated with electric phenomena, and "dissolution". Nevertheless, electrolysis, as a tool to study chemical reactions and obtain pure elements, precedes the coinage of the term and formal description by Faraday.

- 1785 – Martinus van Marum's electrostatic generator was used to reduce tin, zinc, and antimony from their salts using electrolysis.
- 1800 – William Nicholson and Anthony Carlisle (view also Johann Ritter), decomposed water into hydrogen and oxygen.
- 1808 – Potassium (1807), sodium, barium, calcium and magnesium were discovered by Sir Humphry Davy using electrolysis.
- 1821 – Lithium was discovered by William Thomas Brande who obtained it by electrolysis of lithium oxide.
- 1833 – Michael Faraday develops his two laws of electrolysis, and provides a mathematical explanation of his laws.
- 1875 – Paul Émile Lecoq de Boisbaudran discovered gallium using electrolysis.
- 1886 – Fluorine was discovered by Henri Moissan using electrolysis.
- 1886 – Hall–Héroult process developed for making aluminium
- 1890 – Castner–Kellner process developed for making sodium hydroxide

Overview

Electrolysis is the passing of a direct electric current through an ionic substance that is either molten or dissolved in a suitable solvent, producing chemical reactions at the electrodes and separation of materials.

The main components required to achieve electrolysis are:

- An electrolyte: a substance, frequently an ion-conducting polymer that contains free ions, which carry electric current in the electrolyte. If the ions are not mobile, as in a solid salt then electrolysis cannot occur.
- A direct current (DC) electrical supply: provides the energy necessary to create or discharge the ions in the electrolyte. Electric current is carried by electrons in the external circuit.

- Two electrodes: electrical conductors that provide the physical interface between the electrolyte and the electrical circuit that provides the energy.

Electrodes of metal, graphite and semiconductor material are widely used. Choice of suitable electrode depends on chemical reactivity between the electrode and electrolyte and manufacturing cost.

Process of Electrolysis

The key process of electrolysis is the interchange of atoms and ions by the removal or addition of electrons from the external circuit. The desired products of electrolysis are often in a different physical state from the electrolyte and can be removed by some physical processes. For example, in the electrolysis of brine to produce hydrogen and chlorine, the products are gaseous. These gaseous products bubble from the electrolyte and are collected.

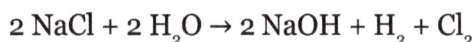

$$2 \, NaCl + 2 \, H_2O \rightarrow 2 \, NaOH + H_2 + Cl_2$$

A liquid containing mobile ions (electrolyte) is produced by:

- Solvation or reaction of an ionic compound with a solvent (such as water) to produce mobile ions

- An ionic compound is fused by heating

An electrical potential is applied across a pair of electrodes immersed in the electrolyte.

Each electrode attracts ions that are of the opposite charge. Positively charged ions (cations) move towards the electron-providing (negative) cathode. Negatively charged ions (anions) move towards the electron-extracting (positive) anode.

In this process electrons are either absorbed or released. Neutral atoms gain or lose electrons and become charged ions that then pass into the electrolyte. The formation of uncharged atoms from ions is called discharging. When an ion gains or loses enough electrons to become uncharged (neutral) atoms, the newly formed atoms separate from the electrolyte. Positive metal ions like Cu^{++} deposit onto the cathode in a layer. The terms for this are electroplating, electrowinning, and electrorefining. When an ion gains or loses electrons without becoming neutral, its electronic charge is altered in the process. In chemistry, the loss of electrons is called Oxidation, while electron gain is called reduction.

Oxidation and Reduction at the Electrodes

Oxidation of ions or neutral molecules occurs at the anode. For example, it is possible to oxidize ferrous ions to ferric ions at the anode:

$$Fe^{2+}(aq) \rightarrow Fe^{3+}(aq) + e^-$$

Reduction of ions or neutral molecules occurs at the cathode.

It is possible to reduce ferricyanide ions to ferrocyanide ions at the cathode:

$$Fe(CN)_6^{3-} + e^- \rightarrow Fe(CN)_6^{4-}$$

Neutral molecules can also react at either of the electrodes. For example: p-Benzoquinone can be reduced to hydroquinone at the cathode:

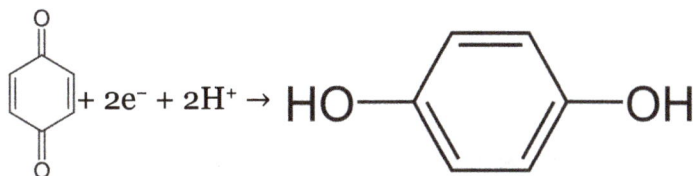

In the last example, H^+ ions (hydrogen ions) also take part in the reaction, and are provided by an acid in the solution, or by the solvent itself (water, methanol etc.). Electrolysis reactions involving H^+ ions are fairly common in acidic solutions. In aqueous alkaline solutions, reactions involving OH^- (hydroxide ions) are common.

Sometimes the solvents themselves (usually water) are oxidized or reduced at the electrodes. It is even possible to have electrolysis involving gases. Such as when using a Gas diffusion electrode.

Energy Changes During Electrolysis

The amount of electrical energy that must be added equals the change in Gibbs free energy of the reaction plus the losses in the system. The losses can (in theory) be arbitrarily close to zero, so the maximum thermodynamic efficiency equals the enthalpy change divided by the free energy change of the reaction. In most cases, the electric input is larger than the enthalpy change of the reaction, so some energy is released in the form of heat. In some cases, for instance, in the electrolysis of steam into hydrogen and oxygen at high temperature, the opposite is true and heat energy is absorbed. This heat is absorbed from the surroundings, and the heating value of the produced hydrogen is higher than the electric input.

Related Techniques

The following techniques are related to electrolysis:

- Electrochemical cells, including the hydrogen fuel cell, use differences in Standard electrode potential to generate an electrical potential that provides useful power. Though related via the interaction of ions and electrodes, electrolysis and the operation of electrochemical cells are quite distinct. However, a chemical cell should *not* be seen as performing *electrolysis in reverse.*

Faraday's Laws of Electrolysis

First Law of Electrolysis

In 1832, Michael Faraday reported that the quantity of elements separated by passing an electric current through a molten or dissolved salt is proportional to the quantity of electric charge passed through the circuit. This became the basis of the first law of electrolysis:

$$m = k \cdot q$$

or

$$m = eQ$$

where; e is known as electrochemical equivalent of the metal deposited or of the gas liberated at the electrode.

Second Law of Electrolysis

Faraday discovered that when the same amount of current is passed through different electrolytes/elements connected in series, the mass of substance liberated/deposited at the electrodes is directly proportional to their equivalent weight.

Industrial Uses

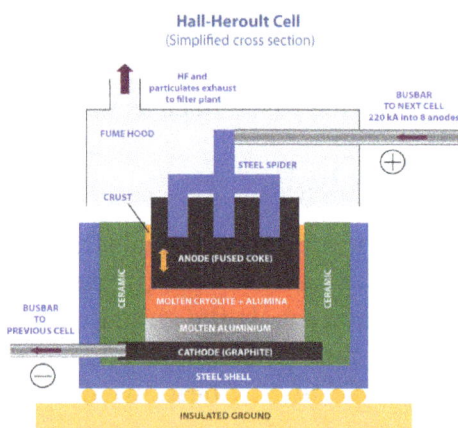

Hall-Heroult process for producing aluminium

- Electrometallurgy is the process of reduction of metals from metallic compounds to obtain the pure form of metal using electrolysis. aluminium, lithium, sodium, potassium, magnesium, calcium, and in some cases copper, are produced in this way.

- Production of chlorine and sodium hydroxide

- Production of sodium chlorate and potassium chlorate

- Production of perfluorinated organic compounds such as trifluoroacetic acid by the process of electrofluorination

- Production of electrolytic copper as a cathode, from refined copper of lower purity as an anode.

Electrolysis has many other uses:

- Production of oxygen for spacecraft and nuclear submarines.

- Production of hydrogen for fuel, using a cheap source of electrical energy.

Electrolysis is also used in the cleaning and preservation of old artifacts. Because the process separates the non-metallic particles from the metallic ones, it is very useful for cleaning a wide variety of metallic objects, from old coins to even larger objects including rusted cast iron cylinder blocks and heads when rebuilding automobile engines. Rust removal from small iron or steel objects by electrolysis can be done in a home workshop using simple materials such as a plastic bucket, tap water, lengths of rebar, washing soda, baling wire, and a battery charger.

Manufacturing Processes

In manufacturing, electrolysis can be used for:

- Electroplating, where a thin film of metal is deposited over a substrate material. Electroplating is used in many industries for either functional or decorative purposes, as in vehicle bodies and nickel coins.

- Electrochemical Machining (ECM), where an electrolytic cathode is used as a shaped tool for removing material by anodic oxidation from a workpiece. ECM is often used as technique for deburring or for etching metal surfaces like tools or knives with a permanent mark or logo.

Competing Half-reactions in Solution Electrolysis

Using a cell containing inert platinum electrodes, electrolysis of aqueous solutions of some salts leads to reduction of the cations (e.g., metal deposition with, e.g., zinc salts) and oxidation of the anions (e.g. evolution of bromine with bromides). However, with salts of some metals (e.g. sodium) hydrogen is evolved at the cathode, and for salts containing some anions (e.g. sulfate SO_4^{2-}) oxygen is evolved at the anode. In both cases this is due to water being reduced to form hydrogen or oxidized to form oxygen. In principle the voltage required to electrolyze a salt solution can be derived from the standard electrode potential for the reactions at the anode and cathode. The standard electrode potential is directly related to the Gibbs free energy, ΔG, for the reactions at each electrode and refers to an electrode with no current flowing. An extract from the table of standard electrode potentials is shown below.

Half-reaction	E° (V)
$Na^+ + e^- \rightleftharpoons Na(s)$	−2.71
$Zn^{2+} + 2e^- \rightleftharpoons Zn(s)$	−0.7618
$2H^+ + 2e^- \rightleftharpoons H_2(g)$	$\equiv 0$
$Br_2(aq) + 2e^- \rightleftharpoons 2Br^-$	+1.0873
$O_2(g) + 4H^+ + 4e^- \rightleftharpoons 2H_2O$	+1.23
$Cl_2(g) + 2e^- \rightleftharpoons 2Cl^-$	+1.36
$S_2O_8^{2-} + 2e^- \rightleftharpoons 2SO_4^{2-}$	+2.07

In terms of electrolysis, this table should be interpreted as follows:

- Oxidized species (often a cation) with a more negative cell potential are more difficult to reduce than oxidized species with a more positive cell potential. For example, it is more difficult to reduce a sodium ion to a sodium metal than it is to reduce a zinc ion to a zinc metal.

- Reduced species (often an anion) with a more positive cell potential are more difficult to oxidize than reduced species with a more negative cell potential. For example, it is more difficult to oxidize sulfate anions than it is to oxidize bromide anions.

Using the Nernst equation the electrode potential can be calculated for a specific concentration of ions, temperature and the number of electrons involved. For pure water (pH 7):

- the electrode potential for the reduction producing hydrogen is −0.41 V

- the electrode potential for the oxidation producing oxygen is +0.82 V.

Comparable figures calculated in a similar way, for 1M zinc bromide, $ZnBr_2$, are −0.76 V for the reduction to Zn metal and +1.10 V for the oxidation producing bromine. The conclusion from these figures is that hydrogen should be produced at the cathode and oxygen at the anode from the electrolysis of water—which is at variance with the experimental observation that zinc metal is deposited and bromine is produced. The explanation is that these calculated potentials only indicate the thermodynamically preferred reaction. In practice many other factors have to be taken into account such as the kinetics of some of the reaction steps involved. These factors together mean that a higher potential is required for the reduction and oxidation of water than predicted, and these are termed overpotentials. Experimentally it is known that overpotentials depend on the design of the cell and the nature of the electrodes.

For the electrolysis of a neutral (pH 7) sodium chloride solution, the reduction of sodium ion is thermodynamically very difficult and water is reduced evolving hydrogen leaving hydroxide ions in solution. At the anode the oxidation of chlorine is observed rather than the oxidation of water since the overpotential for the oxidation of chloride to chlorine is lower than the overpotential for the oxidation of water to oxygen. The hydroxide ions and dissolved chlorine gas react further to form hypochlorous acid. The aqueous solutions resulting from this process is called electrolyzed water and is used as a disinfectant and cleaning agent.

Research Trends

Electrolysis of Carbon Dioxide

The electrochemical reduction or electrocatalytic conversion of CO_2 can produce value-added chemicals such methane, ethylene, ethane, etc. The electrolysis of carbon dioxide gives formate or carbon monoxide, but sometimes more elaborate organic compounds such as ethylene. This technology is under research as a carbon-neutral route to organic compounds.

Electrolysis of Water

Electrolysis of water produces hydrogen.

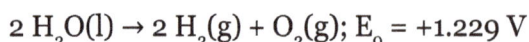

$$2\ H_2O(l) \rightarrow 2\ H_2(g) + O_2(g);\ E_0 = +1.229\ V$$

The energy efficiency of water electrolysis varies widely. The efficiency of an electrolyser is a measure of the enthalpy contained in the hydrogen (to undergo combustion with oxygen, or some other later reaction), compared with the input electrical energy. Heat/enthalpy values for hydrogen are well published in science and engineering texts, as 144 MJ/kg. Note that fuel cells (not electrolysers) cannot utilise this full amount of heat/enthalpy, which has led to some confusion when calculating efficiency values for both types of technology. In the reaction, some energy is lost as heat. Some reports quote efficiencies between 50% and 70% for alkaline electrolysers; however, much higher practical efficiencies are available with the use of PEM (Polymer Electrolyte Membrane electrolysis) and catalytic technology, such as 95% efficiency.

NREL estimated that 1 kg of hydrogen (roughly equivalent to 3 kg, or 4 L, of petroleum in energy terms) could be produced by wind powered electrolysis for between $5.55 in the near term and $2.27 in the long term.

About 4% of hydrogen gas produced worldwide is generated by electrolysis, and normally used onsite. Hydrogen is used for the creation of ammonia for fertilizer via the Haber process, and converting heavy petroleum sources to lighter fractions via hydrocracking.

Carbon/Hydrocarbon Assisted Water Electrolysis

Recently, to reduce the energy input, the utilization of carbon (coal), alcohols (hydrocarbon solution), and organic solution (glycerol, formic acid, ethylene glycol, etc.) with co-electrolysis of water has been proposed as a viable option. The carbon/hydrocarbon assisted electrolysis for hydrogen generation would performs this operation in a single electrochemical reactor. This system energy balance can be required only around 40% electric input with 60% coming from the chemical energy of carbon or hydrocarbon.

Electrocrystallization

A specialized application of electrolysis involves the growth of conductive crystals on one of the electrodes from oxidized or reduced species that are generated in situ. The technique has been used to obtain single crystals of low-dimensional electrical conductors, such as charge-transfer salts.

History

Scientific pioneers of electrolysis include:

- Antoine Lavoisier
- Robert Bunsen
- Humphry Davy
- Michael Faraday
- Paul Héroult
- Svante Arrhenius
- Adolph Wilhelm Hermann Kolbe
- William Nicholson
- Joseph Louis Gay-Lussac
- Alexander von Humboldt

- Johann Wilhelm Hittorf

- Kai Grjotheim

Pioneers of batteries:

- Alessandro Volta

- Gaston Planté

Natural-gas Processing

Natural-gas processing is a complex industrial process designed to clean raw natural gas by separating impurities and various non-methane hydrocarbons and fluids to produce what is known as *pipeline quality* dry natural gas.

Natural-gas processing begins at the well head. The composition of the raw natural gas extracted from producing wells depends on the type, depth, and location of the underground deposit and the geology of the area. Oil and natural gas are often found together in the same reservoir. The natural gas produced from oil wells is generally classified as *associated-dissolved*, meaning that the natural gas is associated with or dissolved in crude oil. Natural gas production absent any association with crude oil is classified as "non-associated." In 2009, 89 percent of U.S. wellhead production of natural gas was non-associated.

Natural-gas processing plants purify raw natural gas by removing common contaminants such as water, carbon dioxide (CO_2) and hydrogen sulfide (H_2S). Some of the substances which contaminate natural gas have economic value and are further processed or sold. A fully operational plant delivers pipeline-quality dry natural gas that can be used as fuel by residential, commercial and industrial consumers.

Types of Raw-natural-gas Wells

Raw natural gas comes primarily from any one of three types of wells: crude oil wells, gas wells, and condensate wells.

Natural gas that comes from crude oil wells is typically called *associated gas*. This gas can have existed as a gas cap above the crude oil in the underground formation, or could have been dissolved in the crude oil.

Natural gas from gas wells and from condensate wells, in which there is little or no crude oil, is called *non-associated gas*. Gas wells typically produce only raw natural gas, while condensate wells produce raw natural gas along with other low molecular weight hydrocarbons. Those that are liquid at ambient conditions (i.e., pentane and heavier) are called *natural gas condensate* (sometimes also called *natural gasoline* or simply *condensate*).

Natural gas is called *sweet gas* when relatively free of hydrogen sulfide; gas that does contain hydrogen sulfide is called *sour gas*. Natural gas, or any other gas mixture, containing significant quantities of hydrogen sulfide, carbon dioxide or similar acidic gases, is called *acid gas*

Raw natural gas can also come from methane deposits in the pores of coal seams, and especially in a more concentrated state of adsorption onto the surface of the coal itself. Such gas is referred to as *coalbed gas* or *coalbed methane* (*coal seam gas* in Australia). Coalbed gas has become an important source of energy in recent decades.

Contaminants in Raw Natural Gas

Raw natural gas typically consists primarily of methane (CH_4), the shortest and lightest hydrocarbon molecule. It also contains varying amounts of:

- Heavier gaseous hydrocarbons: ethane (C_2H_6), propane (C_3H_8), normal butane ($n\text{-}C_4H_{10}$), isobutane ($i\text{-}C_4H_{10}$), pentanes and even higher molecular weight hydrocarbons. When processed and purified into finished by-products, all of these are collectively referred to as Natural Gas Liquids or **NGL**.

- Acid gases: carbon dioxide (CO_2), hydrogen sulfide (H_2S) and mercaptans such as methanethiol (CH_3SH) and ethanethiol (C_2H_5SH).

- Other gases: nitrogen (N_2) and helium (He).

- Water: water vapor and liquid water. Also dissolved salts and dissolved gases (acids).

- Liquid hydrocarbons: perhaps some natural-gas condensate (also referred to as *casing-head gasoline* or *natural gasoline*) and/or crude oil.

- Mercury: very small amounts of mercury primarily in elemental form, but chlorides and other species are possibly present.

- Naturally occurring radioactive material (NORM): natural gas may contain radon, and the produced water may contain dissolved traces of radium, which can accumulate within piping and processing equipment. This can render piping and equipment radioactive over time.

The raw natural gas must be purified to meet the quality standards specified by the major pipeline transmission and distribution companies. Those quality standards vary from pipeline to pipeline and are usually a function of a pipeline system's design and the markets that it serves. In general, the standards specify that the natural gas:

- Be within a specific range of heating value (caloric value). For example, in the United States, it should be about 1035 ± 5% BTU per cubic foot of gas at 1 atmosphere and 60°F (41 MJ ± 5% per cubic metre of gas at 1 atmosphere and 15.6°C).

- Be delivered at or above a specified hydrocarbon dew point temperature (below which some of the hydrocarbons in the gas might condense at pipeline pressure forming liquid slugs that could damage the pipeline).

- Dew-point adjustment serves the reduction of the concentration of water and heavy hydrocarbons in natural gas to such an extent that no condensation occurs during the ensuing transport in the pipelines

- Be free of particulate solids and liquid water to prevent erosion, corrosion or other damage to the pipeline.

- Be dehydrated of water vapor sufficiently to prevent the formation of methane hydrates within the gas processing plant or subsequently within the sales gas transmission pipeline. A typical water content specification in the U.S. is that gas must contain no more than seven pounds of water per million standard cubic feet (MMSCF) of gas.

- Contain no more than trace amounts of components such as hydrogen sulfide, carbon dioxide, mercaptans, and nitrogen. The most common specification for hydrogen sulfide content is 0.25 grain H_2S per 100 cubic feet of gas, or approximately 4 ppm. Specifications for CO_2 typically limit the content to no more than two or three percent.

- Maintain mercury at less than detectable limits (approximately 0.001 ppb by volume) primarily to avoid damaging equipment in the gas processing plant or the pipeline transmission system from mercury amalgamation and embrittlement of aluminum and other metals.

Description of a Natural-gas Processing Plant

There are a great many ways in which to configure the various unit processes used in the processing of raw natural gas. The block flow diagram below is a generalized, typical configuration for the processing of raw natural gas from non-associated gas wells. It shows how raw natural gas is processed into sales gas pipelined to the end user markets. It also shows how processing of the raw natural gas yields these byproducts:

- Natural-gas condensate

- Sulfur

- Ethane

- Natural-gas liquids (NGL): propane, butanes and C_5+ (which is the commonly used term for pentanes plus higher molecular weight hydrocarbons)

Raw natural gas is commonly collected from a group of adjacent wells and is first processed at that collection point for removal of free liquid water and natural gas condensate. The condensate is usually then transported to an oil refinery and the water is disposed of as wastewater.

The raw gas is then pipelined to a gas processing plant where the initial purification is usually the removal of acid gases (hydrogen sulfide and carbon dioxide). There are many processes that are available for that purpose as shown in the flow diagram, but amine treating is the process that was historically used. However, due to a range of performance and environmental constraints of the amine process, a newer technology based on the use of polymeric membranes to separate the carbon dioxide and hydrogen sulfide from the natural gas stream has gained increasing acceptance. Membranes are attractive since no reagents are consumed.

The acid gases, if present, are removed by membrane or amine treating can then be routed into a sulfur recovery unit which converts the hydrogen sulfide in the acid gas into either elemental sulfur or sulfuric acid. Of the processes available for these conversions, the Claus process is by far the most well known for recovering elemental sulfur, whereas the conventional Contact process and the WSA (Wet sulfuric acid process) are the most used technologies for recovering sulfuric acid.

The residual gas from the Claus process is commonly called *tail gas* and that gas is then processed in a tail gas treating unit (TGTU) to recover and recycle residual sulfur-containing compounds back into the Claus unit. Again, as shown in the flow diagram, there are a number of processes available for treating the Claus unit tail gas and for that purpose a WSA process is also very suitable since it can work autothermally on tail gases.

The next step in the gas processing plant is to remove water vapor from the gas using either the regenerable absorption in liquid triethylene glycol (TEG), commonly referred to as glycol dehydration, deliquescent chloride desiccants, and or a Pressure Swing Adsorption (PSA) unit which is regenerable adsorption using a solid adsorbent. Other newer processes like membranes may also be considered.

Mercury is then removed by using adsorption processes such as activated carbon or regenerable molecular sieves.

Although not common, nitrogen is sometimes removed and rejected using one of the three processes indicated on the flow diagram:

- Cryogenic process (Nitrogen Rejection Unit), using low temperature distillation. This process can be modified to also recover helium, if desired.

- Absorption process, using lean oil or a special solvent as the absorbent.

- Adsorption process, using activated carbon or molecular sieves as the adsorbent. This process may have limited applicability because it is said to incur the loss of butanes and heavier hydrocarbons.

The next step is to recover the natural gas liquids (NGL) for which most large, modern gas processing plants use another cryogenic low temperature distillation process involving expansion of the gas through a turbo-expander followed by distillation in a demethanizing fractionating column. Some gas processing plants use lean oil absorption process rather than the cryogenic turbo-expander process.

The residue gas from the NGL recovery section is the final, purified sales gas which is pipelined to the end-user markets.

The recovered NGL stream is sometimes processed through a fractionation train consisting of three distillation towers in series: a deethanizer, a depropanizer and a debutanizer. The overhead product from the deethanizer is ethane and the bottoms are fed to the depropanizer. The overhead product from the depropanizer is propane and the bottoms are fed to the debutanizer. The overhead product from the debutanizer is a mixture of normal and iso-butane, and the bottoms product is a C_5+ mixture. The recovered streams of propane, butanes and C_5+ may be "sweetened" in a Merox process unit to convert undesirable mercaptans into disulfides and, along with the recovered ethane, are the final NGL by-products from the gas processing plant. Currently, most cryogenic plants do not include fractionation for economic reasons, and the NGL stream is instead transported as a mixed product to standalone fractionation complexes located near refineries or chemical plants that use the components for feedstock. In case laying pipeline is not possible for geographical reason,or the distance between source and consumer exceed 3000 km, natural gas is then transported by ship as LNG (liquefied natural gas) and again converted into its gaseous state in the vicinity of the consumer.

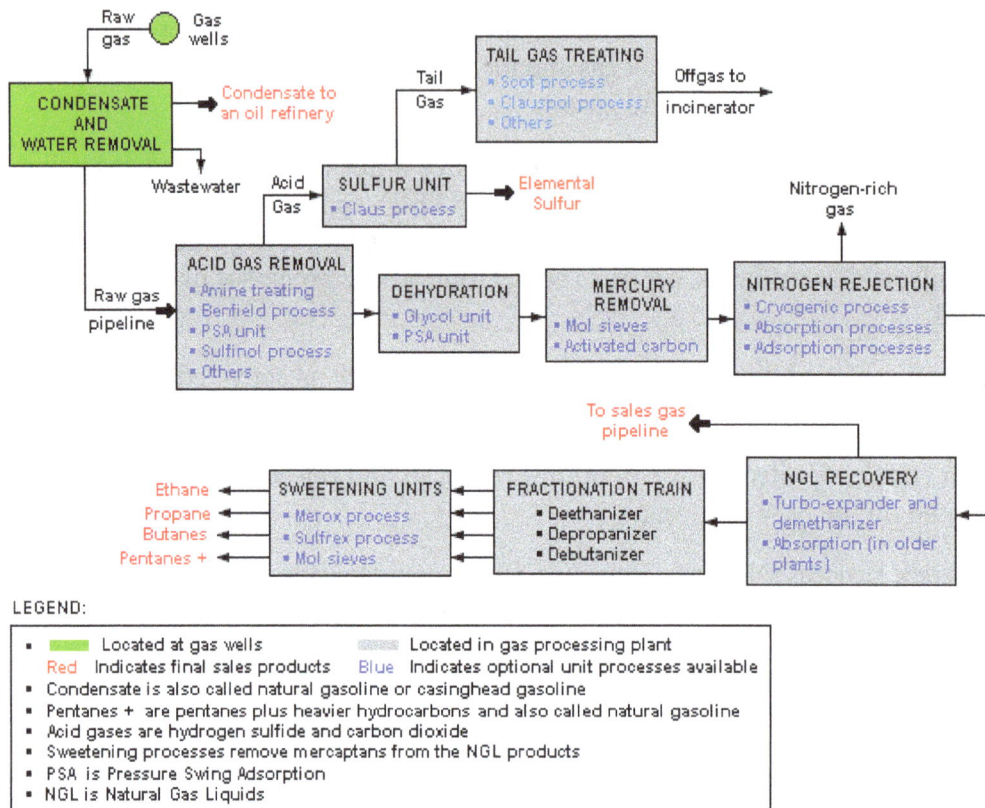

Helium Recovery

If the gas contains significant helium content, the helium may be recovered by fractional distillation. Natural gas may contain as much as 7% helium, and is the commercial source of the noble gas. For instance, the Hugoton Gas Field in Kansas and Oklahoma in the United States contains concentrations of helium from 0.3% to 1.9%, which is separated out as a valuable byproduct.

Consumption

Natural gas consumption patterns, across nations, vary based on access. Countries with large reserves tend to handle the raw-material natural gas more generously, while countries with scarce or lacking resources tend to be more economical. Despite the considerable findings, the predicted availability of the natural-gas reserves has hardly changed.

Applications of Natural Gas

- Fuel for industrial heating and desiccation process
- Fuel for the operation of public and industrial power stations

- Household fuel for cooking, heating and providing hot water

- Fuel for environmentally friendly compressed or liquid natural gas vehicles

- Raw material for chemical synthesis

- Raw material for large-scale fuel production using gas-to-liquid (GTL) process (e.g. to produce sulphur-and aromatic-free diesel with low-emission combustion)

Electrosynthesis

Electrosynthesis in chemistry is the synthesis of chemical compounds in an electrochemical cell. The main advantage of electrosynthesis over an ordinary redox reaction is avoidance of the potential wasteful other half-reaction and the ability to precisely tune the required potential. Electrosynthesis is actively studied as a science and also has many industrial applications. Electrooxidation is studied not only for synthesis but also for efficient removal of certain harmful organic compounds in wastewater.

Experimental Setup

The basic setup in electrosynthesis is a galvanic cell, a potentiostat and two electrodes. Good electrosynthetic conditions use a solvent and electrolyte combination that minimizes electrical resistance. Protic conditions often use alcohol-water or dioxane-water solvent mixtures with an electrolyte such as a soluble salt, acid or base. Aprotic conditions often use an organic solvent such as acetonitrile or dichloromethane with electrolytes such as lithium perchlorate or tetrabutylammonium acetate. Electrodes are selected which provide favorable electron transfer properties towards the substrate while maximizing the activation energy for side reactions. This activation energy is often related to an overpotential of a competing reaction. For example, in aqueous conditions the competing reactions in the cell are the formation of oxygen at the anode and hydrogen at the cathode. In this case a graphite anode and lead cathode could be used effectively because of their high overpotentials for oxygen and hydrogen formation respectively. Many other materials can be used as electrodes. Other examples include platinum, magnesium, mercury (as a liquid pool in the reactor), stainless steel or reticulated vitreous carbon. Some reactions use a sacrificial electrode is used which is consumed during the reaction like zinc or lead. The two basic cell types are undivided cell or divided cell type. In divided cells the cathode and anode chambers are separated with a semiporous membrane. Common membrane materials include sintered glass, porous porcelain, polytetrafluoroethene or polypropylene. The purpose of the divided cell is to permit the diffusion of ions while restricting the flow of the products and reactants. This is important when unwanted side reactions are possible. An example of a reaction requiring a divided cell is the reduction of nitrobenzene to phenylhydroxylamine, where the latter chemical is susceptible to oxidation at the anode.

Reactions

Organic oxidations take place at the anode with initial formation of radical cations as reactive intermediates. Compounds are reduced at the cathode to radical anions. The initial reaction takes place at the surface of the electrode and then the intermediates diffuse into the solution where they participate in secondary reactions.

The yield of an electrosynthesis is expressed both in terms the chemical yield and current efficiency. Current efficiency is the ratio of Coulombs consumed in forming the products to the total number of Coulombs passed through the cell. Side reactions decrease the current efficiency.

The potential drop between the electrodes determines the rate constant of the reaction. Electrosynthesis is carried out with either constant potential or constant current. The reason one chooses one over the other is due to a trade off of ease of experimental conditions versus current efficiency. Constant potential uses current more efficiently because the current in the cell decreases with time due to the depletion of the substrate around the working electrode (stirring is usually necessary to decrease the diffusion layer around the electrode). This is not the case under constant current conditions however. Instead as the substrate's concentration decreases the potential across the cell increases in order to maintain the fixed reaction rate. This consumes current in side reactions produced outside the target voltage.

Anodic Oxidations

- The most well-known electrosynthesis is the Kolbe electrolysis, in which two carboxylic acids decarboxylate, and the remaining structures bond together:

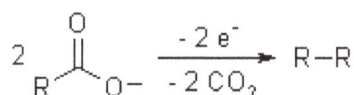

- A variation is called the non-Kolbe reaction when a heteroatom (nitrogen or oxygen) is present at the α-position. The intermediate oxonium ion is trapped by a nucleophile usually solvent.

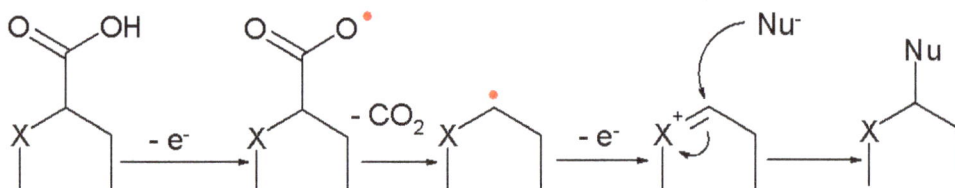

- Amides can be oxidized to N-acyliminium ions, which can be captured by various nucleophiles, for example:

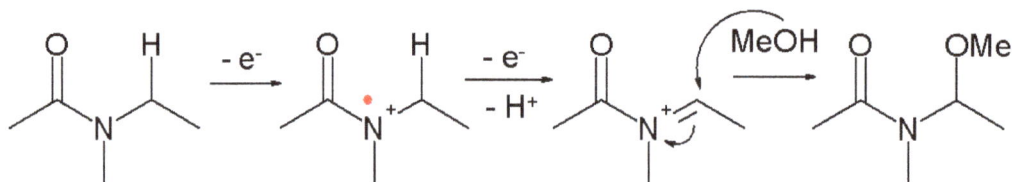

This reaction type is called a Shono oxidation. An example is the α-methoxylation of *N*-carbomethoxypyrrolidine

- Oxidation of a carbanion can lead to a coupling reaction for instance in the electrosynthesis of the tetramethyl ester of ethanetetracarboxylic acid from the corresponding malonate ester

- α-amino acids form nitriles and carbon dioxide via oxidative decarboxylation at AgO anodes (the latter is formed *in-situ* by oxidation of Ag_2O):

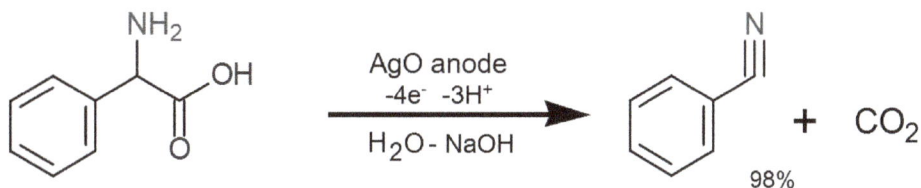

- Cyanoacetic acid from cathodic reduction of carbon dioxide and anodic oxidation of acetonitrile.

Cathodic Reductions

- In the Markó–Lam deoxygenation, an alcohol could be almost instantaneously deoxygenated by electroreducing their toluate ester.

- The cathodic hydroisomerization of activated olefins is applied industrially in the synthesis of adiponitrile from 2 equivalents of acrylonitrile:

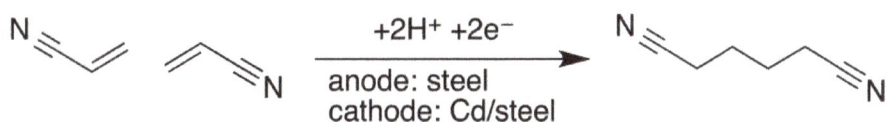

- The cathodic reduction of arene compounds to the 1,4-dihydro derivatives is similar to a Birch reduction. Examples from industry are the reduction of phthalic acid:

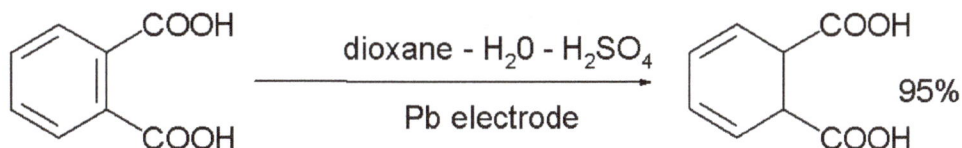

and the reduction of 2-methoxynaphthalene:

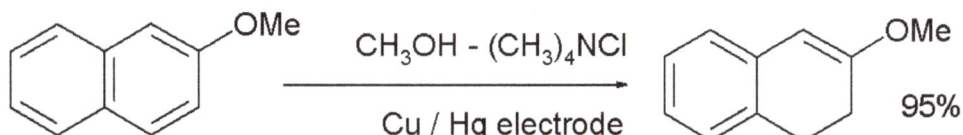

- The Tafel rearrangement, named for Julius Tafel, was at one time an important method for the synthesis of certain hydrocarbons from alkylated ethyl acetoacetate, a reaction accompanied by the rearrangement reaction of the alkyl group:

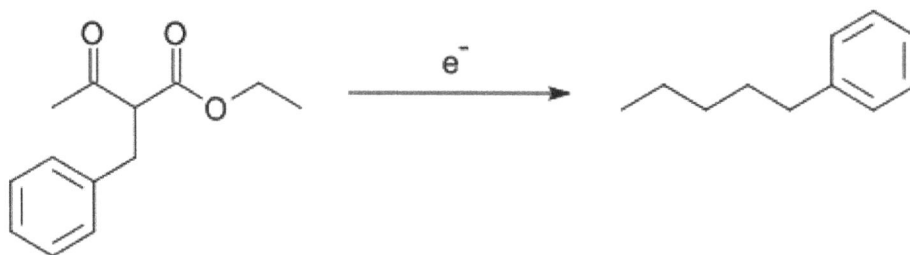

- The cathodic reduction of a nitrile to a primary amine in a divided cell:

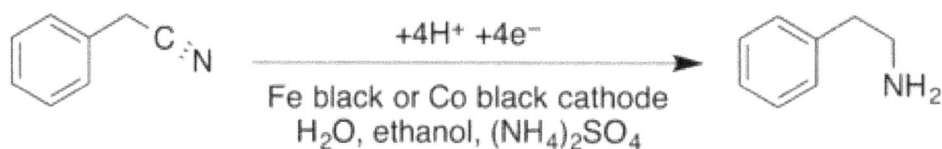

- Cathodic reduction of a nitroalkene can give the oxime in good yield. At higher negative reduction potentials, the nitroalkene can be reduced further, giving the primary amine but with lower yield.

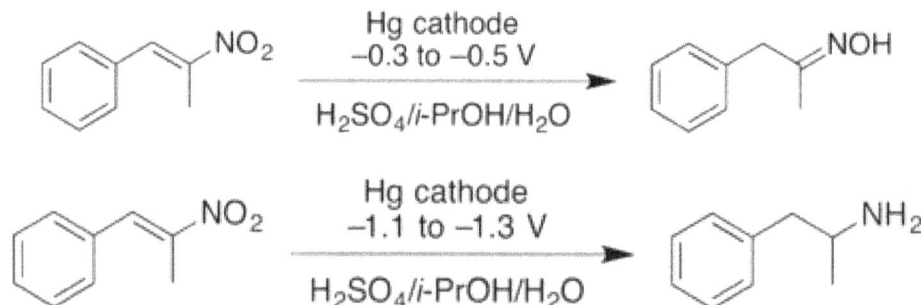

- An electrochemical carboxylation of a para-isobutylbenzyl chloride to Ibuprofen is promoted under supercritical carbon dioxide.

- Cathodic reduction of a carboxylic acid (oxalic acid) to an aldehyde (glyoxylic acid, shows as the rare aldehyde form) in a divided cell:

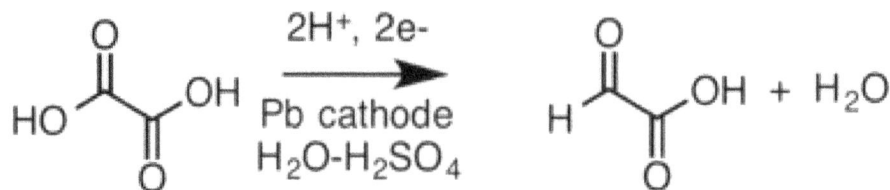

- An electrocatalysis by a copper complex helps reduce carbon dioxide to oxalic acid; this conversion uses carbon dioxide as a feedstock to generate oxalic acid.

Electrofluorination

In organofluorine chemistry, many perfluorinated compounds are prepared by electrochemical synthesis, which is conducted in liquid HF at voltages near 5–6 V using Ni anodes. The method was invented in the 1930s. Amines, alcohols, carboxylic acids, and sulfonic acids are converted to the perfluorinated derivatives using this technology. A solution or suspension of the hydrocarbon in hydrogen fluoride is electrolyzed at 5–6 V to produce high yields of the perfluorinated product.

Galvanization

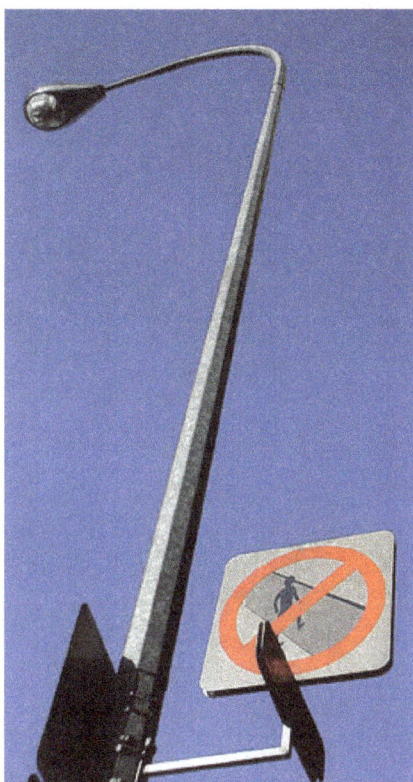

A street lamp in Singapore showing the characteristic spangle of hot-dip galvanization.

Galvanization or galvanisation (or galvanizing as it is most commonly called in that industry) is the process of applying a protective zinc coating to steel or iron, to prevent rusting. The most common method is hot-dip galvanizing, in which parts are submerged in a bath of molten zinc. Galvanizing protects in three ways:

- It forms a coating of zinc which, when intact, prevents corrosive substances from reaching the underlying steel or iron.

- The zinc serves as a sacrificial anode so that even if the coating is scratched, the exposed steel will still be protected by the remaining zinc.

- The zinc protects its base metal by corroding before iron. For better results, application of chromates over zinc is also seen as an industrial trend.

Etymology

The earliest known example of galvanizing of iron, encountered by Europeans is found on 17th-century Indian armor in the Royal Armouries Museum collection. It was named in English via French from the name of Italian scientist Luigi Galvani. Originally, galvanizing was the administration of electric shocks, in the 19th century also termed Faradism. This sense is the origin of the meaning of the metaphorical use of the verb 'galvanize', as in 'galvanize into action', or to stimulate a complacent person or group to take action. The term galvanizing has largely come to be associated with zinc coatings, to the exclusion of other metals. Galvanic paint, a precursor to hot-dip galvanizing, was patented by Stanislas Sorel, of Paris in December, 1837.

Galvanized nails.

Methods

Galvanized surface with visible spangle

Hot-dip galvanizing deposits a thick robust layer of zinc iron alloys on the surface of a steel item. In the case of automobile bodies, where additional decorative coatings of paint will be applied, a thinner form of galvanizing is applied by electrogalvanizing. The hot-dip process generally does not reduce strength on a measurable scale, with the exception of high-strength steels (>1100 MPa) where hydrogen embrittlement can become a problem. This is a consideration for the manufacture of wire rope and other highly stressed products. The protection provided by hot-dip galvanizing is insufficient for products that will be constantly exposed to corrosive materials such as acids. For these applications, more expensive stainless steel is preferred. Some nails made today are galvanized. Nonetheless, electroplating is used on its own for many outdoor applications because it is cheaper than hot-dip zinc coating and looks good when new. Another reason not to use hot-dip zinc coating is that for bolts and nuts size M10 (US 3/8") or smaller, the thick hot-dipped coating fills in too much of the threads, which reduces strength (because the dimension of the steel prior to coating must be reduced for the fasteners to fit together). This means that for cars, bicycles, and many other light mechanical products, the alternative to electroplating bolts and nuts is not hot-dip zinc coating, but making the bolts and nuts from stainless steel.

The size of crystallites in galvanized coatings is a visible and aesthetic feature, known as "spangle". By varying the number of particles added for heterogeneous nucleation and the rate of cooling in a hot-dip process, the spangle can be adjusted from an apparently uniform surface (crystallites too small to see with the naked eye) to grains several centimetres wide. Visible crystallites are rare in other engineering materials.

Thermal diffusion galvanizing, or Sherardizing, provides a zinc diffusion coating on iron- or copper-based materials. Parts and zinc powder are tumbled in a sealed rotating drum. Around 300°C, zinc will diffuse into the substrate to form a zinc alloy. The preparation of the goods can be carried out by shot blasting. The process is also known as dry galvanizing, because no liquids are involved; however, no danger of hydrogen embrittlement of the goods exists. The dull-grey crystal structure of the zinc diffusion coating has a good adhesion to paint, powder coatings, or rubber. It is a preferred method for coating small, complex-shaped metals, and for smoothing rough surfaces on items formed with powder metal.

Eventual Corrosion

Rusted corrugated steel roof

Although galvanizing will inhibit attack of the underlying steel, rusting will be inevitable, after some decades, especially if exposed to acidic conditions. For example, corrugated iron sheet roofing will start to degrade within a few years despite the protective action of the zinc coating. Marine and salty environments also lower the lifetime of galvanized iron because the high electrical conductivity of sea water increases the rate of corrosion primarily through converting the solid zinc to soluble zinc chloride which simply washes away. Galvanized car frames exemplify this; they corrode much quicker in cold environments due to road salt, though they will last longer than unprotected steel. Galvanized steel can last for many decades if other means are maintained, such as paint coatings and additional sacrificial anodes. The rate of corrosion in non-salty environments is mainly due to levels of sulfur dioxide in the air. In the most benign natural environments, such as inland low population areas, galvanized steel can last without rust for over 100 years.

Galvanized Piping

In the early 20th century, galvanized piping replaced cast iron and lead in cold-water plumbing. Typically, galvanized piping rusts from the inside out, building up plaques on the inside of the piping, causing both water pressure problems and eventual pipe failure. These plaques can flake off, leading to visible impurities in water and a slight metallic taste. The life expectancy of such piping is about 70 years, but it may vary by region due to impurities in the water supply and the proximity of electrical grids for which interior piping acts as a pathway (the flow of electricity can accelerate chemical corrosion). Pipe longevity also depends on the thickness of zinc in the original galvanizing, which ranges on a scale from G40 to G210, and whether the pipe was galvanized on both the inside and outside, or just the outside. Since World War II, copper and plastic piping have replaced galvanized piping for interior drinking water service, but galvanized steel pipes are still used in outdoor applications requiring steel's superior mechanical strength.

This lends some truth to the urban myth that water purity in outdoor water faucets is lower, but the actual impurities (iron, zinc, calcium) are harmless.

The presence of galvanized piping detracts from the appraised value of housing stock because piping can fail, increasing the risk of water damage. Galvanized piping will eventually need to be replaced if housing stock is to outlast a 50 to 70 year life expectancy, and some jurisdictions require galvanized piping to be replaced before sale. One option to extend the life expectancy of existing galvanized piping is to line it with an epoxy resin.

Galvanized Construction Steel

This is the most common use for galvanizing, and hundreds of thousands of tonnes are galvanized annually worldwide. In developed countries most larger cities have several galvanizing factories, and many items of steel manufacture are galvanized for protection. Typically these include: street furniture, building frameworks, balconies, verandahs, staircases, ladders, walkways and more. SGCC hot dip galvanized steel is also used for making steel frames as a basic construction material for steel frame buildings.

Chemical Garden

A chemical garden is an experiment in chemistry normally performed by adding metal salts such as copper sulfate or cobalt(II) chloride to an aqueous solution of sodium silicate (otherwise known as waterglass). This results in growth of plant-like forms in minutes to hours.

A chemical garden

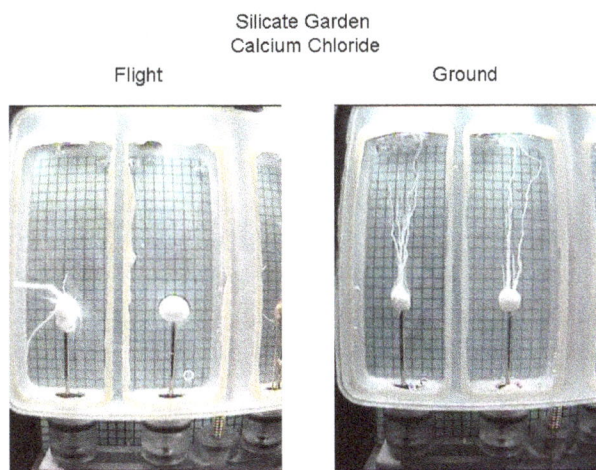

Silicate Garden
Calcium Chloride

Flight   Ground

Comparison of chemical gardens grown by NASA scientists on the International Space Station (left) and on the ground (right)

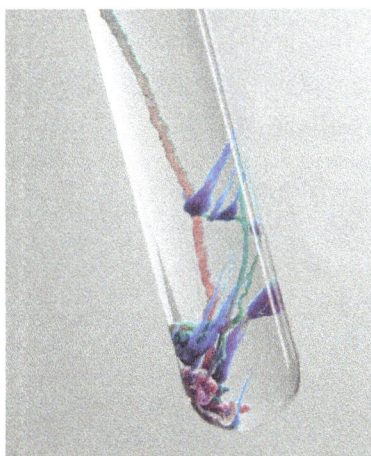

Cobalt(II) chloride

The chemical garden was first observed and described by Johann Rudolf Glauber in 1646. In its original form, the chemical garden involved the introduction of ferrous chloride ($FeCl_2$) crystals into a solution of potassium silicate (K_2SiO_3).

Process

The chemical garden relies on most transition metal silicates being insoluble in water and coloured.

A metal salt such as cobalt chloride will start to dissolve in the water. It will then form insoluble cobalt silicate by a double decomposition reaction (anion metathesis). This cobalt silicate is a semipermeable membrane. Because the ionic strength of the cobalt solution inside the membrane is higher than the sodium silicate solution which forms the bulk of the tank contents, osmotic effects will increase the pressure within the membrane. This will cause the membrane to tear, forming a hole. The cobalt cations will react with the silicate anions at this tear so forming new solid. In this way growths will form in the tanks; these will be coloured (according to the metal) and may look like plants. The crystals formed from this experiment will grow upwards, since the pressure at the bottom of the tank is higher than the pressure closer to the top of the tank, therefore forcing the crystals to grow upwards.

The upward direction of growth depends on the density of the fluid inside the semi-permeable membrane being lower than that of the surrounding waterglass solution. If one uses a very dense fluid inside the membrane, the growth is downward. For example, a fresh, green solution of tri-valent chromium sulfate or chloride refuses to crystallise without slowly changing into the violet form, even if boiled till it concentrates into a tarry mass. That tar, if suspended in the waterglass solution, forms downward twig-like growths because all the fluid inside the membrane is too dense to float and thereby to exert upward pressure. The concentration of sodium silicate becomes important in growth rate. Although any concentration works, a ratio of 2ml of water to 3ml of sodium silicate works best.

After the growth has ceased sodium silicate solution can be removed by continuous addition of water at a very slow rate. This will prolong the life of garden. In one specific experimental variation a single tube can be obtained

Common Salts Used

Common salts used in a chemical garden include:

- Aluminium potassium sulfate: White
- Copper(II) sulfate: Blue
- Chromium(III) chloride: Green
- Nickel(II) sulfate: Green
- Iron(II) sulfate: Green
- Iron(III) chloride: Orange
- Cobalt(II) chloride: Purple
- Calcium chloride: White
- Zinc sulfate: White

Practical Uses

While at the chemical garden may appear to be primarily a toy, some serious work has been done on the subject. For instance this chemistry is related to the setting of Portland cement, hydrother-mal vents and during the corrosion of steel surfaces tubes can be formed. A chemical garden helps one to understand the nature of that chemical substance.

The nature of the growth also is useful in understanding classes of related behaviour seen in fluids separated by membranes. In various ways it resembles the growth of spikes or blobs of ice extrud-ed above the freezing surface of still water, the patterns of growth of gum drying as it drips from wounds in trees such as Eucalyptus sap, and the way molten wax forms twig-like growths, either dripping from a candle, or floating up through cool water.

Atom Economy

Atom economy (atom efficiency) is the conversion efficiency of a chemical process in terms of all atoms involved and the desired products produced. Atom economy is an important concept of green chemistry philosophy, and one of the most widely used metrics for measuring the "greenness" of a process or synthesis.

Atom economy can be written as:

$$\text{atom economy} = \frac{\text{molecular mass of desired product}}{\text{molecular mass of all reactants}} \times 100\%$$

By the conservation of mass, the total molecular mass of the reactants is the same as the total molecular mass of the products. In an ideal chemical process, the amount of starting materials or reactants equals the amount of all products generated and no atom is wasted. However, in some processes, some of the consumed reactant atoms do not become part of the intended products. This can be a concern for raw materials that have a high cost or due to economic and environmental costs of disposal of the waste.

Atom economy is a different concern than chemical yield, because a high-yielding process can still result in substantial byproducts. Examples include the Cannizzaro reaction, where approximately 50% of the reactant aldehyde becomes the other oxidation state of the target, the Wittig reaction, where high-mass phosphorus reagents are used but ultimately become waste, and the Gabriel synthesis, which produces a stoichiometric quantity of phthalic acid

If the desired product has an enantiomer the reaction needs to be sufficiently stereoselective even when atom economy is 100%. A Diels-Alder reaction is an example of a potentially very atom efficient reaction that also can be chemo-, regio-, diastereo- and enantioselective. Catalytic hydrogenation comes the closest to being an ideal reaction that is extensively practiced both industrially and academically.

Atom economy can also be adjusted if a pendant group is recoverable, for example Evans auxiliary groups. However, if this can be avoided it is more desirable, as recovery processes will never be 100%. Atom economy can be improved upon by careful selection of starting materials and a catalyst system.

Poor atom economy is common in fine chemicals or pharmaceuticals synthesis, and especially in research, where the aim to readily and reliably produce a wide range of complex compounds leads to the use of versatile and dependable, but poorly atom-economical reactions. For example, synthesis of an alcohol is readily accomplished by reduction of an ester with lithium aluminum hydride, but the reaction necessarily produces a voluminous floc of aluminum salts, which have to be separated from the product alcohol and disposed of. The cost of such hazardous material disposal can be considerable. Catalytic hydrogenolysis of an ester is the analogous reaction with a high atom economy, but it requires catalyst optimization, is a much slower reaction and is not applicable universally.

Creating Reactions Utilizing Atom Economy

It is fundamental in chemical reactions of the form A+B→ C+D that two products are necessarily generated though product C may have been the desired one. That being the case, D is considered a byproduct. As it is a significant goal of green chemistry to maximize the efficiency of the reactants and minimize the production of waste, D must either be found to have use, be eliminated or be as insignificant and innocuous as possible. With the new equation of the form A+B→C, the first step in making chemical manufacturing more efficient is the use of reactions that resemble simple addition reactions with the only other additions being catalytic materials.

Asymmetric Hydrogenation

Asymmetric hydrogenation is a chemical reaction that adds two atoms of hydrogen preferentially to one of two faces of an unsaturated substrate molecule, such as an alkene or ketone. The selectivity derives from the manner that the substrate binds to the chiral catalysts. In jargon, this binding transmits spatial information (what chemists refer to as chirality) from the catalyst to the target, favoring the product as a single enantiomer. This enzyme-like selectivity is particularly applied to bioactive products such as pharmaceutical agents and agrochemicals.

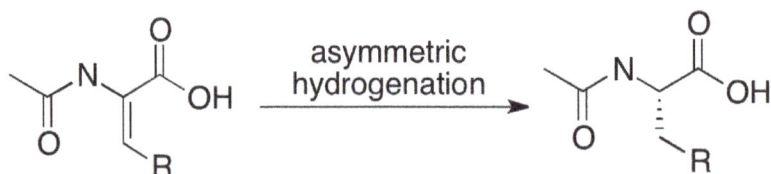

History

In 1956 a heterogeneous catalyst made of palladium deposited on silk was shown to effect asymmetric hydrogenation. Later, in 1968, the groups of William Knowles and Leopold Horner independently published the examples of asymmetric hydrogenation using a homogeneous catalysts. While exhibiting only modest enantiomeric excesses, these early reactions demonstrated feasibility. By 1972, enantiomeric excess of 90% was achieved, and the first industrial synthesis of the Parkinson's drug L-DOPA commenced using this technology.

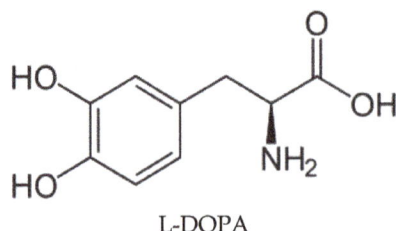

L-DOPA

The field of asymmetric hydrogenation continued to experience a number of notable advances. Henri Kagan developed DIOP, an easily prepared C_2-symmetric diphosphine that gave high ee's in certain reactions. Ryōji Noyori introduced the ruthenium-based catalysts for the asymmetric hydrogenated polar substrates, such as ketones and aldehydes. The introduction of P,N ligands then further expanded the scope of the C_2-symmetric ligands, although they are not fundamentally superior to chiral ligands lacking rotational symmetry. Today, asymmetric hydrogenation is a routine methodology in laboratory and industrial scale organic chemistry.

The importance of asymmetric hydrogenation was recognized by the 2001 Nobel Prize in Chemistry awarded to William Standish Knowles and Ryōji Noyori.

Mechanism

Two major mechanisms have been proposed for catalytic hydrogenation with rhodium complexes: the unsaturated mechanism and the dihydride mechanism. While distinguishing between the two mechanisms is difficult, the difference between the two for asymmetric hydrogenation is relatively unimportant since both converge to a common intermediate before any stereochemical information is transferred to the product molecule.

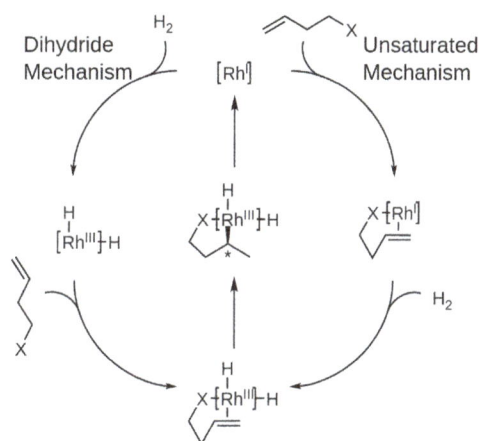

Proposed mechanisms for asymmetric hydrogenation

The preference for producing one enantiomer instead of another in these reactions is often explained in terms of steric interactions between the ligand and the prochiral substrate. Consideration of these interactions has led to the development of quadrant diagrams where "blocked" areas are denoted with a shaded box, while "open" areas are left unfilled. In the modeled reaction, large groups on an incoming olefin will tend to orient to fill the open areas of the diagram, while smaller groups will be directed to the blocked areas and hydrogen delivery will then occur to the back face of the olefin, fixing the stereochemistry. Note that only part of the chiral phosphine ligand is shown for the sake of clarity.

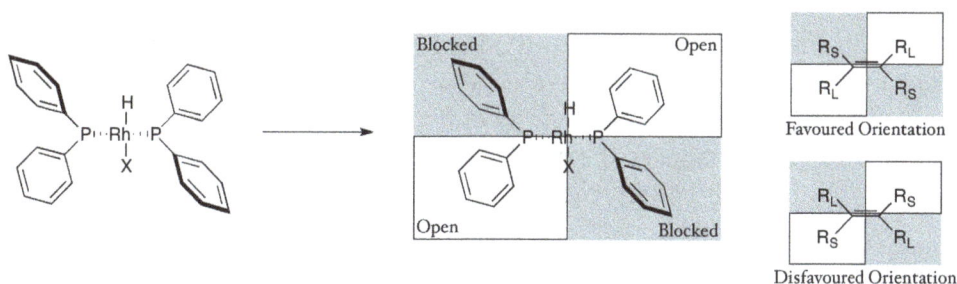

Quadrant model for asymmetric hydrogenation

Metals

Platinum-group Metals

Rhodium, the first metal to be used in a homogeneous asymmetric hydrogenation, continues to be widely used. Targets for asymmetric hydrogenation with rhodium generally require a coordinating group close to the olefin. While this requirement is a limitation, many classes of substrates possess such functionalization, e.g. unsaturated amides.

The Noyori asymmetric hydrogenation is based on ruthenium. Subsequent work has expanded upon Noyori's original catalyst template, leading to the inclusion of traditionally difficult substrates like *t*-butyl ketones and 1-tetralones as viable substrates for hydrogenation with ruthenium catalysts. Transfer hydrogenation based on the Ru and TsDPEN has also enjoyed commercial success.

Iridium catalysts are useful for a number of "non-traditional" substrates for which good catalysts had not been found with Ru and Rh. Unfunctionalized olefins are the archetypal case, but other examples including ketones exist. A common difficulty with iridium-based catalyst is their tendency to trimerize in solution. The use of a $BArF_4^-$ anions has proven to be the most widely applicable solution to the aggregation problem. Other strategies to enhance catalyst stability include the addition of an additional coordinating arm to the chiral ligand, increasing the steric bulk of the ligand, using a dendrimeric ligand, increasing the rigidity of the ligand, immobilizing the ligand, and using heterobimetallic systems (with iridium as one of the metals).

Base Metals

Iron is a popular research target for many catalytic processes, owing largely to its low cost and low toxicity relative to other transition metals. Asymmetric hydrogenation methods using iron have been realized, although in terms of rates and selectivity, they are inferior to catalysts based on precious metals. In some cases, structurally ill-defined nanoparticles have proven to be the active species *in situ* and the modest selectivity observed may result from their uncontrolled geometries.

Ligand Classes

Phosphine Ligands

Chiral phosphine ligands are the source of chirality in most asymmetric hydrogenation catalysts. Of these the BINAP ligand is perhaps the best-known, as a result of its Nobel Prize-winning application in the Noyori asymmetric hydrogenation.

Chiral phosphine ligands can be generally classified as mono- or bidentate. They can be further classified according to the location of the stereogenic centre – phosphorus vs the organic substituents. Ligands with a C_2 symmetry element have been particularly popular, in part because the presence of such an element reduces the possible binding conformations of a substrate to a metal-ligand complex dramatically (often resulting in exceptional enantioselectivity).

Chiral Monophosphine, Monodentate Ligands

Monophosphine-type ligands were among the first to appear in asymmetric hydrogenation, e.g., the ligand CAMP. Continued research into these types of ligands has explored both P-alkyl and P-heteroatom bonded ligands, with P-heteroatom ligands like the phosphites and phosphoramidites generally achieving more impressive results. Structural classes of ligands that have been successful include those based on the binapthyl structure of MonoPHOS or the spiro ring system of SiPHOS. Notably, these monodentate ligands can be used in combination with each other to achieve a synergistic improvement in enantioselectivity; something that is not possible with the diphosphine ligands.

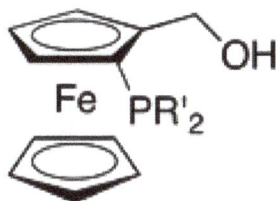
A ferrocene derivative

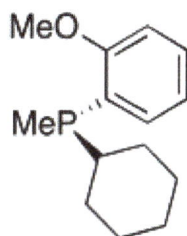
The CAMP ligand

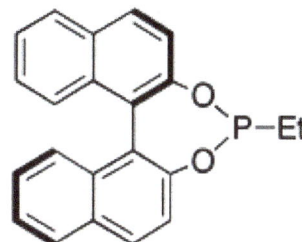
A BINOL derivative

Chiral Diphosphine Ligands

The diphosphine ligands have received considerably more attention than the monophosphines and, perhaps as a consequence, have a much longer list of achievement. This class includes the first ligand to achieve high selectivity (DIOP), the first ligand to be used in industrial asymmetric synthesis (DIPAMP) and what is likely the best known chiral ligand (BINAP). Chiral diphosphine ligands are now ubiquitous in asymmetric hydrogenation.

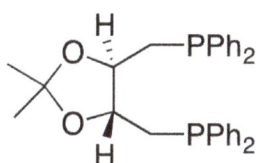
(R,R)-DIOP

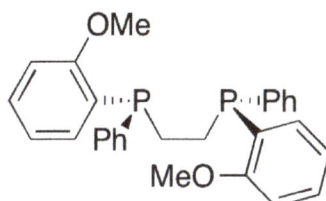
(R,R)-DIPAMP

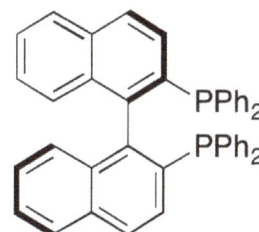
(R)-BINAP

Historically important diphosphine ligands

P,N and P,O Ligands

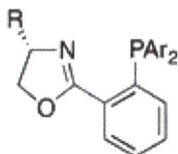
Generic PHOX ligand architecture

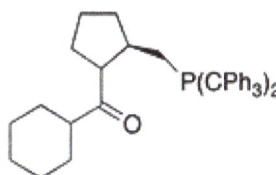
Effective ligand for various asymmetric-hydrogenation processes

The use of P,N ligands in asymmetric hydrogenation can be traced to the C_2 symmetric bisoxazoline ligand. However, these symmetric ligands were soon superseded by mono(oxazoline) ligands whose lack of C_2 symmetry has in no way limits their efficacy in asymmetric catalysis. Such ligands generally consist of an achiral nitrogen-containing heterocycle that is functionalized with a pendant phosphorus-containing arm, although both the exact nature of the heterocycle and the chemical environment phosphorus center has varied widely. No single structure has emerged as consistently effective with a broad range of substrates, although certain privileged structures (like the phosphine-oxazoline or PHOX architecture) have been established. Moreover, within a narrowly defined substrate class the performance of metallic complexes with chiral P,N ligands can closely approach perfect conversion and selectivity in systems otherwise very difficult to target. Certain complexes derived from chelating P-O ligands have shown promising results in the hydrogenation of α,β-unsaturated ketones and esters.

NHC Ligands

Simple N-heterocyclic carbene (NHC)-based ligands have proven impractical for asymmetrical hydrogenation.

Some C,N ligands combine an NHC with a chiral oxazoline to give a chelating ligand. NHC-based ligands of the first type have been generated as large libraries from the reaction of smaller libraries of individual NHCs and oxazolines. NHC-based catalysts featuring a bulky seven-membered metallocycle on iridium have been applied to the catalytic hydrogenation of unfunctionalized olefins and vinyl ether alcohols with conversions and ee's in the high 80s or 90s. The same system has been applied to the synthesis of a number of aldol, vicinal dimethyl and deoxypolyketide motifs, and to the deoxypolyketides themselves.

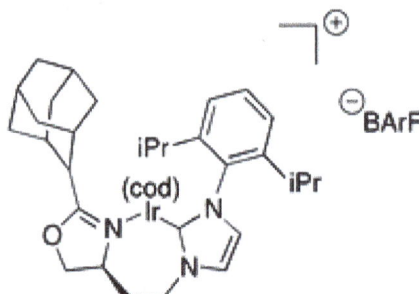

Catalyst developed by Burgess for asymmetric hydrogenation

C_2-symmetric NHCs have shown themselves to be highly useful ligands for the asymmetric hydrogenation.

Acyclic Substrates

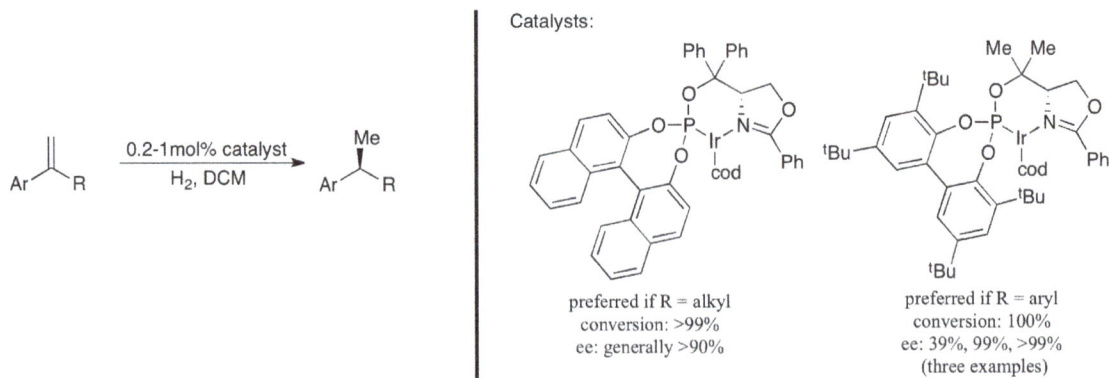

Example of asymmetric hydrogenation of unfunctionalized olefins

Acyclic unsaturated substrates (olefins, ketones, enamines imines) represents the most common prochiral substrates. Substrates that are particularly amenable to asymmetric hydrogenation often feature a polar functional group adjacent to the site to be hydrogenates. In the absence of this functional) group, catalysis often results in low ee's. For unfunctionalized olefins, iridium with P,N-based ligands) have proven successful catalysts. Catalyst utility within this category is unusually narrow; consequently, many different categories of solved and unsolved catalytic problems have developed. 1,1-disubstituted, 1,2-diaryl trisubstituted, 1,1,2-trialkyl and tetrasubstituted olefins represent classes that have been investigated separately, and even within these classes variations may exist that make different solutions optimal.

Conversely to the case of olefins, asymmetric hydrogenation of enamines has favoured diphosphine-type ligands; excellent results have been achieved with both iridium- and rhodium-based systems. However, even the best systems often suffer from low ee's and a lack of generality. Certain pyrrolidine-derived enamines of aromatic ketones are amenable to asymmetrically hydrogenation with cationic rhodium(I) phosphonite systems, and I_2 and acetic acid system with ee values usually above 90% and potentially as high as 99.9%. A similar system using iridium(I) and a very closely related phosphoramidite ligand is effective for the asymmetric hydrogenation of pyrrolidine-type enamines where the double bond was inside the ring: in other words, of dihydropyrroles. In both cases, the enantioselectivity dropped substantially when the ring size was increased from five to six.

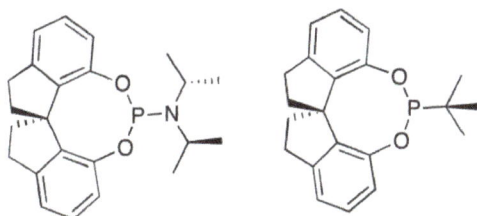

Chiral phosphoramidite and phosphonite ligands used in the asymmetric hydrogenation of enamines.

Imines and Ketones

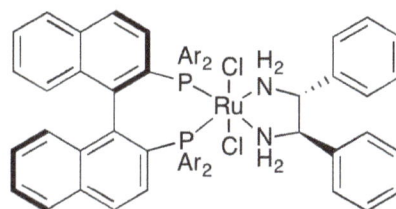

Archetype of Noyori's catalysts for asymmetric hydrogenation of ketones

Ketones and imines are related functional groups, and effective technologies for the asymmetric hydrogenation of each are also closely related. Of these, Noyori's ruthenium-chiral diphosphine-diamine system is perhaps one of the best known. It can be employed in conjunction with a wide range of phosphines and amines (where the amine may be, but need not be, chiral) and can be easily adjusted for an optimal match with the target substrate, generally achieving enantiomeric excesses (ee's) above 90%.

For carbonyl and imine substrates, end-on, η^1 coordination can compete with η^2 mode. For η^1-bound substrates, the hydrogen-accepting carbon is removed from the catalyst and resists hydrogenation.

Iridium/P,N ligand-based systems are also commonly used for the asymmetric hydrogenation of ketones and imines. For example, a consistent system for benzylic aryl imines uses the P,N ligand SIPHOX in conjunction with iridium(I) in a cationic complex to achieve asymmetric hydrogenation with ee >90%. One of the most efficient and effective catalysts ever developed for the asymmetric hydrogenation of ketones, with a turnover number (TON) up to 4,550,000 and ee up to 99.9%, uses another iridium(I) system with a closely related tridentate ligand.

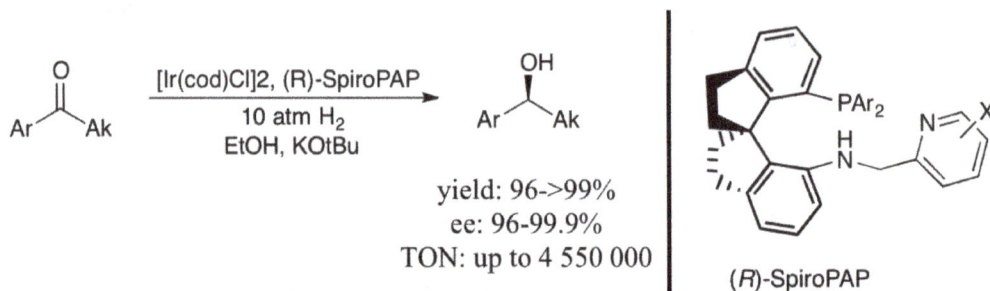

yield: 96->99%
ee: 96-99.9%
TON: up to 4 550 000

(R)-SpiroPAP

Highly effective system for the asymmetric hydrogenation of ketones

Despite their similarities, the two functional groups are not identical; there are many areas where they diverge significantly. One of these is in the asymmetric hydrogenation of N-unfunctionalized imines to give primary amines. Such species can be difficult to selectively reduce because they tend to exist in complex equilibria of imine and enamine tautomers, as well as (E) and (Z) isomers. One approach to this problem has been to use ketimines as their hydrochloride salt and rely on the steric properties of the adjacent alkyl or aryl groups to allow the catalyst to differentiate between the two enantiotopic faces of the ketimine.

Aromatic Substrates

The asymmetric hydrogenation of aromatic (especially heteroaromatic), substrates is a very active field of ongoing research. Catalysts in this field must contend with a number of complicating factors, including the tendency of highly stable aromatic compounds to resist hydrogenation, the potential coordinating (and therefore catalyst-poisoning) abilities of both substrate and product, and the great diversity in substitution patterns that may be present on any one aromatic ring. Of these substrates the most consistent success has been seen with nitrogen-containing heterocycles, where the aromatic ring is often activated either by protonation or by further functionalization of the nitrogen (generally with an electron-withdrawing protecting group). Such strategies are less applicable to oxygen- and sulfur-containing heterocycles, since they are both less basic and less nucleophilic; this additional difficulty may help to explain why few effective methods exist for their asymmetric hydrogenation.

Quinolines, Isoquinolines and Quinoxalines

Two systems exist for the asymmetric hydrogenation of 2-substituted quinolines with isolated yields generally greater than 80% and ee values generally greater than 90%. The first is an iridium(I)/chiral phosphine/I_2 system, first reported by Zhou et al. While the first chiral phosphine used in this system was MeOBiPhep, newer iterations have focused on improving the performance of this ligand. To this end, systems use phosphines (or related ligands) with improved air stability, recyclability, ease of preparation, lower catalyst loading and the potential role of achiral phosphine additives. As of October 2012 no mechanism appears to have been proposed, although both the necessity of I_2 or a halogen surrogate and the possible role of the heteroaromatic N in assisting reactivity have been documented.

The second is an organocatalytic transfer hydrogenation system based on Hantzsch esters and a chiral Brønsted acid. In this case, the authors envision a mechanism where the isoquinoline is alternately protonated in an activating step, then reduced by conjugate addition of hydride from the Hantzsch ester.

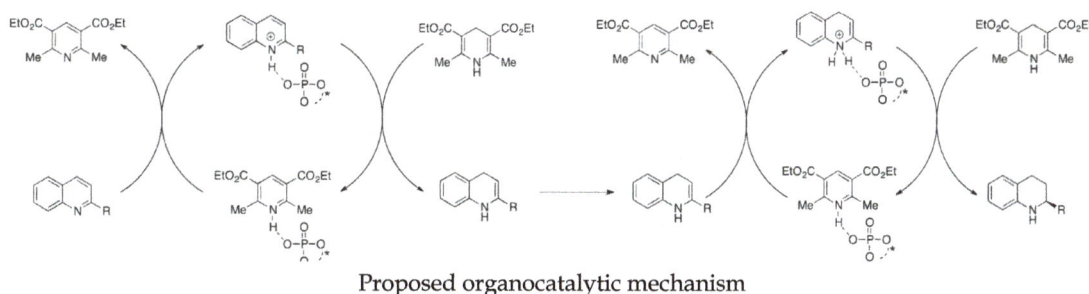

Proposed organocatalytic mechanism

Much of the asymmetric hydrogenation chemistry of quinoxalines is closely related to that of the structurally similar quinolines. Effective (and efficient) results can be obtained with an Ir(I)/phophinite/I_2 system and a Hantzsh ester-based organocatalytic system, both of which are similar to the systems discussed earlier with regards to quinolines.

Pyridines

Pyridines are highly variable substrates for asymmetric reduction (even compared to other heteroaromatics), in that five carbon centers are available for differential substitution on the initial ring. As of October 2012 no method seems to exist that can control all five, although at least one reasonably general method exists.

The most-general method of asymmetric pyridine hydrogenation is actually a heterogeneous method, where asymmetry is generated from a chiral oxazolidinone bound to the C2 position of the pyridine. Hydrogenating such functionalized pyridines over a number of different heterogeneous metal catalysts gave the corresponding piperidine with the substituents at C3, C4, and C5 positions in an all-*cis* geometry, in high yield and excellent enantioselectivity. The oxazolidinone auxiliary is also conveniently cleaved under the hydrogenation conditions.

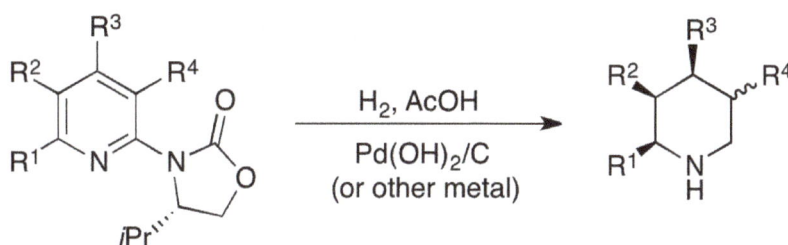

Asymmetric hydrogenation of pyridines with heterogeneous catalyst

Methods designed specifically for 2-substituted pyridine hydrogenation can involve asymmetric systems developed for related substrates like 2-substituted quinolines and quinoxalines. For example, an iridium(I)\chiral phosphine\I_2 system is effective in the asymmetric hydrogenation of activated (alkylated) 2-pyridiniums or certain cyclohexanone-fused pyridines. Similarly, chiral Brønsted acid catalysis with a Hantzsh ester as a hydride source is effective for some 2-alkyl pyridines with additional activating substitution.

Indoles

The asymmetric hydrogenation of indoles initially focused on N-protected indoles, where the protecting group could serve both to activate the heterocycle to hydrogenation and as a secondary coordination site for the metal. Later work allowed unprotected indoles to be targeted through Brønsted acid activation of the indole.

In the initial report on asymmetric indole hydrogenation, N-acetyl 2-substituted indoles could be protected with high yields and ee of 87-95%. 3-substituted indoles were less successful, with hydrolysis of the protecting group outcompeting the hydrogenation of the indole. Switching to an N-tosyl protecting group inhibited the hydrolysis reaction and allowed both 2- and 3-substituted indoles to be hydrogenated in high yield and ee. The problem with both methods, however, is that N-acetyl and N-tosyl groups require harsh cleavage conditions that might be incompatible with complex substrates. Using an easily cleaved N-Boc group would relieve this problem, and highly effective methods for the asymmetric hydrogenation of such indoles (both 2- and 3-substituted) were soon developed.

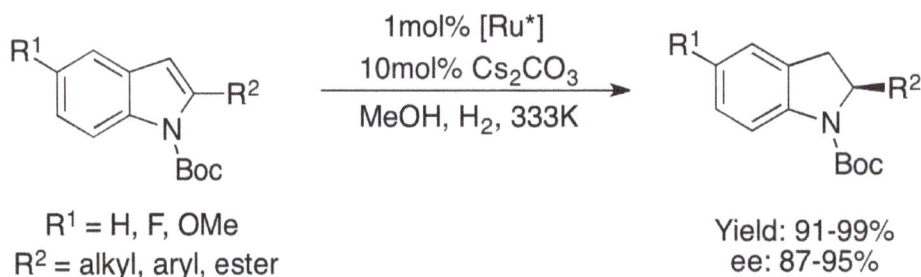

1mol% [Ru*]
10mol% Cs_2CO_3
MeOH, H_2, 333K

R^1 = H, F, OMe
R^2 = alkyl, aryl, ester

Yield: 91-99%
ee: 87-95%

Method for asymmetric hydrogenation of boc-protected indoles

Despite these advances in the asymmetric hydrogenation of protected indoles, considerable operational simplicity can be gained by removing the protecting group altogether. This has been achieved with catalytic systems utilizing Brønsted acids to activate the indole. The initial system used a Pd(T-FA)$_2$/H8-BINAP system to achieve the enantioselective cis-hydrogenation of 2,3- and 2-substituted indoles with high yield and excellent ee. A similar process, where sequential Friedel-Crafts alkylation and asymmetric hydrogenation occur in one pot, allow asymmetric 2,3-substituted indolines to be selectively prepared from 2-substituted indoles in similarly high yields and ee.

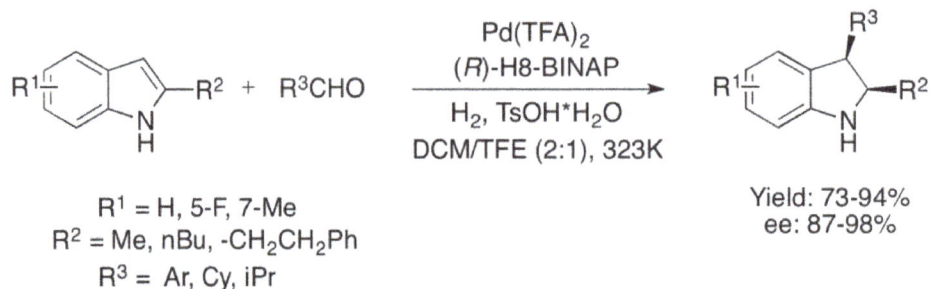

R^3CHO

Pd(TFA)$_2$
(R)-H8-BINAP
H_2, TsOH*H_2O
DCM/TFE (2:1), 323K

R^1 = H, 5-F, 7-Me
R^2 = Me, nBu, -CH$_2$CH$_2$Ph
R^3 = Ar, Cy, iPr

Yield: 73-94%
ee: 87-98%

Sequential alkylation and asymmetric hydrogenation of 2-substituted indoles

A promising organocatalytic method for the asymmetric hydrogenation of 2,3-substituted indoles utilizing a chiral Lewis base also exists, although the observed ee's are not quite equivalent to those of the metal-based hydrogenations.

Pyrroles

Achieving complete conversion of pyrroles to pyrrolidines by asymmetric hydrogenation has so far proven difficult, with partial-hydrogenation products often being observed. Complete enantioselective reduction is possible, with the outcome depending on both the starting substrate and the method.

The asymmetric hydrogenation of 2,3,5-substituted pyrroles was achieved by the recognition that such substrates bear the same substitution pattern as 2-substituted indoles, and an asymmetric hydrogenation system that is effective for one of these substrates might be effective for both. Such an analysis led to the development of a ruthenium(I)/phosphine/amine base system for 2,3,5-substituted N-Boc pyrroles that can give either dihydro or tetrahydropyrroles (pyrrolidines), depending on the nature of the pyrrole substituents. An all-phenyl substitution pattern leads to dihydropyrroles in very high yield (>96%) and essentially perfect enantioselectivity. Access to the fully hydrogenated, all-*cis* dihydropyrrole may then be accessible through diastereoselective heterogeneous hydrogenation. Alkyl substitution may lead to either the dihydro or tetrahydropyrrole, although the yields (>70%) and enantioselectivities (often >90%) generally remain high. The regioselectivity in both cases appears to be governed by sterics, with the less-substituted double being preferentially hydrogenated.

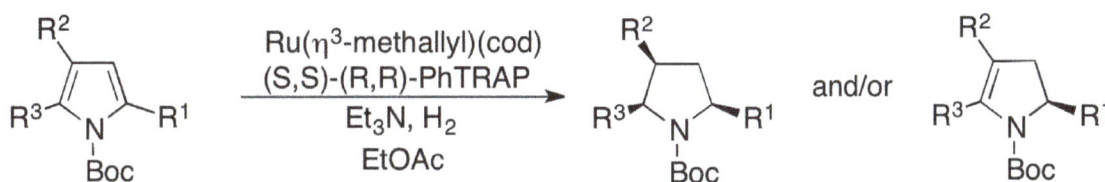

The asymmetric hydrogenation of 2,3,5-substituted N-Boc pyrroles

Unprotected 2,5-pyrroles may also be hydrogenated asymmetrically by a Brønsted acid/Pd(II)/chiral phosphine-catalyzed method, to give the corresponding 2,5-disubstituted 1-pyrrolines in roughly 70-80% yield and 80-90% ee.

Oxygen-containing Heterocycles

The asymmetric hydrogenation of furans and benzofurans has so far proven challenging. Some Ru-NHC complex catalyze asymmetric hydrogenations of benzofurans and furans. with high levels of enantioinduction.

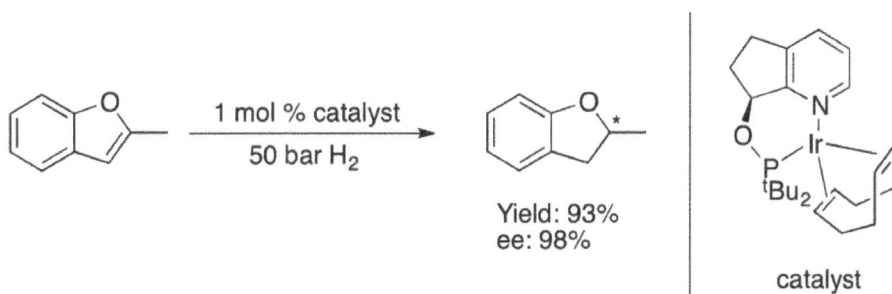

The asymmetric hydrogenation of furans and benzofurans

Sulfur-containing Heterocycles

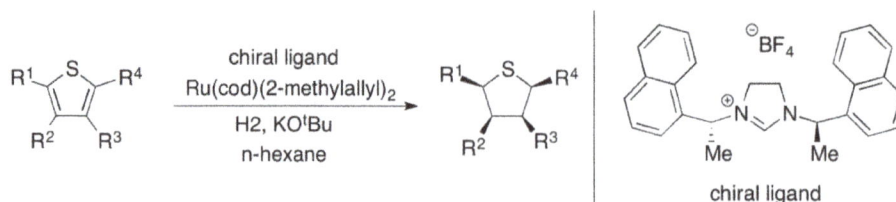

The asymmetric hydrogenation of thiophenes and benzothiophenes

As is the case with oxygen-containing heterocycles, the asymmetric hydrogenation of compounds where sulfur is part of the initial unsaturated pi-bonding system so far appears to be limited to thiophenes and benzothiophenes. The key approach to the asymmetric hydrogenation of these heterocycles involves a ruthenium(II) catalyst and chiral, C_2 symmetric N-heterocyclic carbene (NHC). Interestingly, this system appears to possess superb selectivity (ee > 90%) and perfect diastereoselectivity (all *cis*) if the substrate has a fused (or directly bound) phenyl ring but yields only racemic product in all other tested cases.

Heterogeneous Catalysis

Research into asymmetric hydrogenation with heterogeneous catalysts has generally focused on three areas. The oldest, dating back to the first asymmetric hydrogenation with palladium deposited on a silk support, involves modifying a metal surface with a chiral molecule, usually one that can be harvested from nature. Alternatively, researchers have used various techniques to attempt to immobilize what would otherwise be homogeneous catalysts on heterogeneous supports or have used synthetic organic ligands and metal sources to build chiral metal-organic frameworks (MOFs).

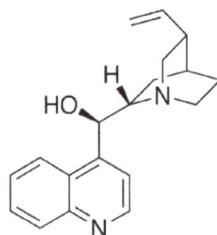

Cinchonidine, one of the cinchona alkaloids

The greatest successes in chiral modification of metal surfaces have come from the use of cinchona alkaloids, though numerous other classes of natural products have been evaluated. These alkaloids have been shown to enhance the rate of substrate hydrogenation by 10–100 times, such that less than one molecule of cinchona alkaloid is needed for every reactive site on the metal and, in fact, the presence of too much of the chiral modifier can cause a decrease in the enantioselectivity of the reaction.

An alternative technique and one that allows more control over the structural and electronic properties of active catalytic sites is the immobilization of catalysts that have been developed for homogeneous catalyis on a heterogeneous support. Covalent bonding of the catalyst to a polymer or other solid support is perhaps most common, though immobilization of the catalyst may also be achieved by adsorption onto a surface, ion exchange, or even physical encapsulation. One drawback of this approach is the potential for the proximity of the support to change the behaviour of the catalyst, lowering the enantioselectivity of the reaction. To avoid this, the catalyst is often bound to the support by a long linker though cases are known where the proximity of the support can actually enhance the performance of the catalyst.

The final approach involves the construction of MOFs that incorporate chiral reaction sites from a number of different components, potentially including chiral and achiral organic ligands, structural metal ions, catalytically active metal ions, and/or preassembled catalytically active organometallic cores. This field is relatively new, and few examples exist of chiral asymmetric hydrogenation using these frameworks. One of these was reported in 2003, when a heterogeneous catalyst was reported that included structural zirconium, catalytically active ruthenium, and a BINAP-derived phosphonate as both chiral ligand and structural linker. As little as 0.005 mol% of this catalyst proved sufficient to achieve the asymmetric hydrogenation of aryl ketones, though the usual conditions featured 0.1 mol % of catalyst and resulted in an enantiomeric excess of 90.6–99.2%.

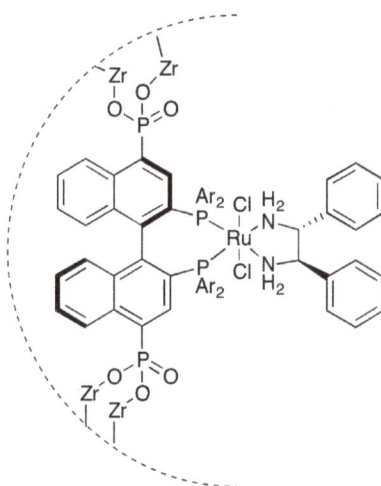

The active site of a heterogeneous zirconium phosphonate catalyst for asymmetric hydrogenation

Industrial Applications

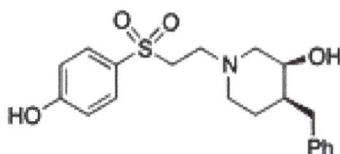

(S,S)-Ro 67-8867

Knowles' research into asymmetric hydrogenation and its application to the production scale synthesis of L-Dopa gave asymmetric hydrogenation a strong start in the industrial world. More recently, a 2001 review indicated that asymmetric hydrogenation accounted for 50% of production scale, 90% of pilot scale, and 74% of bench scale catalytic, enantioselective processes in industry, with the caveat that asymmetric catalytic methods in general were not yet widely used.

The success of asymmetric hydrogenation in industry can be seen in a number of specific cases where the replacement of kinetic resolution based methods has resulted in substantial improvements in the process's efficiency. For example, Roche's Catalysis Group was able to achieve the synthesis of (S,S)-Ro 67-8867 in 53% overall yield, a dramatic increase above the 3.5% that was achieved in the resolution based synthesis. Roche's synthesis of mibefradil was likewise improved by replacing resolution with asymmetric hydrogenation, reducing the step count by three and increasing the yield of a key intermediate to 80% from the original 70%.

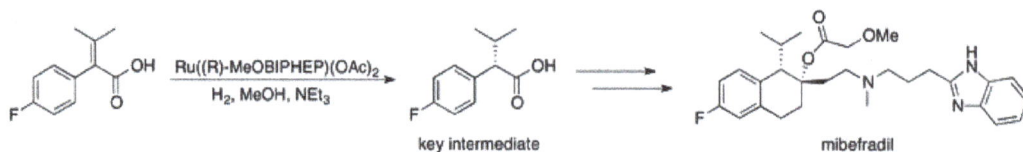

Asymmetric hydrogenation in the industrial synthesis of mibefradil

Supercritical Hydrolysis

Supercritical hydrolysis is a chemical engineering process in which water in the supercritical state can be employed to achieve a variety of reactions within seconds. To cope with the extremely short times of reaction on an industrial scale, the process should be continuous. This continuity enables the ratio of the amount of water to the other reactant to be less than unity which minimizes the energy needed to heat the water above 374 C, the critical point. Application of the process to biomass provides simple sugars in near quantitative yield by supercritical hydrolysis of the constituent polysaccharides. The phenolic polymer components of the biomass, usually exemplified by lignins, are converted into a water-insoluble liquid mixture of low molecular phenols.

A private company, Renmatix, based in King of Prussia, PA, has developed a supercritical hydrolysis technology to convert a range of non-food biomass feedstocks into cellulosic sugars for application in biochemicals and biofuels. It has a demonstration facility in Georgia, currently capable of processing three dry tons of hardwood biomass into cellulosic sugar daily. In Australia, a government-sponsored entity called Licella, is similarly transforming sawdust. Both processes require high ratios of water to the amount of feedstock. This energy profligacy can be avoided by the use of a plastic-type extruder through which the solid, but wet, biomass is conveyed to a small inductively heated reaction zone as shown by Xtrudx Technologies Inc of Seattle.

Supercritical hydrolysis can be considered a broadly applicable green chemistry process that utilizes water simultaneously as a heat transfer agent, a solvent, a reactant, a source of hydrogen and as a char-reduction component.

References

- R. J. D. Tilley (2004). Understanding solids: the science of materials. John Wiley and Sons. pp. 281–. ISBN 978-0-470-85276-7. Retrieved 22 October 2011.

- Grimshaw, James (2000). Electrochemical Reactions and Mechanisms in Organic Chemistry. Amsterdam: Elsevier Science. pp. 1–7, 282, & 310. ISBN 9780444720078.

- Blaser, Hans-Ulrich; Federsel, Hans-Jürgen, eds. (2010). Asymmetric Catalysis on Industrial Scale. Weinheim: Wiley-VCH. pp. 13–16. doi:10.1002/9783527630639. ISBN 978-3-527-63063-9.

- Jacobsen, E.N.; Pfaltz, Andreas; Yamamato, H., eds. (1999). Comprehensive Asymmetric Catalysis. Berlin;

New York: Springer. pp. 1443–1445. ISBN 3-540-64336-2.

- H. Ju et al., Electro-catalytic conversion of ethanol in solid electrolyte cells for distributed hydrogen generation, Electrochimica Acta 212 (2016) 744–757 doi:10.1016/j.electacta.2016.07.062

- S. Giddey et al., Low emission hydrogen generation through carbon assisted electrolysis International Journal of Hydrogen Energy 40 (2015) 70-74 doi:10.1016/j.ijhydene.2014.11.033

- Wysocki, Jędrzej; Ortega, Nuria; Glorius, Frank (2014). "Asymmetric Hydrogenation of Disubstituted Furans". Angewandte Chemie International Edition. 53 (33): 8751. doi:10.1002/anie.201310985.

- J. Qiao, et al., A review of catalysts for the electroreduction of carbon dioxide to produce low-carbon fuels, Chem.Soc.Rev., 2014, 43 , 631-675.

- Appel, A. M. et al. "Frontiers, Opportunities, and Challenges in Biochemical and Chemical Catalysis of CO_2 Fixation", Chem. Rev. 2013, vol. 113, 6621-6658. doi:10.1021/cr300463y

- Carmo, M; Fritz D; Mergel J; Stolten D (2013). "A comprehensive review on PEM water electrolysis". Journal of Hydrogen Energy. doi:10.1016/j.ijhydene.2013.01.151.

- Franzke, A.; Pfaltz, A. (2011). "Zwitterionic Iridium Complexes with P,N-Ligands as Catalysts for the Asymmetric Hydrogenation of Alkenes". Chemistry: A European Journal. 17 (15): 4131. doi:10.1002/chem.201003314.

Green Chemistry: Problem Solving Approaches

The problem solving approaches considered in green chemistry are alternatives assessment, advanced oxidation process, California green chemistry initiative and International Conference on Green Chemistry. Alternatives assessment is a problem solving approach that is used in environmental technologies and policies. The topics discussed in the chapter are of great importance to broaden the existing knowledge on green chemistry.

Alternatives Assessment

Alternatives assessment or alternatives analysis is a problem-solving approach used in environmental design, technology, and policy. It aims to minimize environmental harm by comparing multiple potential solutions in the context of a specific problem, design goal, or policy objective. It is intended to inform decision-making in situations with many possible courses of action, a wide range of variables to consider, and significant degrees of uncertainty. Alternatives assessment was originally developed as a robust way to guide precautionary action and avoid paralysis by analysis; authors such as O'Brien have presented alternatives assessment as an approach that is complementary to risk assessment, the dominant decision-making approach in environmental policy. Likewise, Ashford has described the similar concept of *technology options analysis* as a way to generate innovative solutions to the problems of industrial pollution more effectively than through risk-based regulation.

Alternatives assessment is practiced in a variety of settings, including but not limited to green chemistry, sustainable design, supply-chain chemicals management, and chemicals policy. One prominent application area for alternatives assessment is the substitution of hazardous chemicals with safer alternatives, also known as chemical alternatives assessment.

Methodology

Generally, alternatives assessment involves considering a number of possible options to achieve a specific objective, and applying a principled comparative analysis. The objective is usually to improve the environmental performance or safety of a specific product, material, process, or other activity. Potential alternatives considered in the analysis may include different chemical substances, materials, technologies, methods of use, or even extensive redesign to enable new ways of achieving the objective while avoiding the problem. Understanding the consequences of each available option is central to the process and goals of alternatives assessment, because this helps avoid decisions that substitute one problem with another (unknown) problem. The comparative analysis can involve any number of criteria for evaluating options, and these are typically focused on environmental health and sustainability.

There is no single protocol that dictates how options should be identified, evaluated, and compared in an alternatives assessment. Rather, a number of different alternatives assessment *frameworks* exist, which serve to structure decision-making and to enable systematic consideration of the key factors. Jacobs and colleagues identify six major components of alternatives assessment: evaluation of hazard, exposure, life cycle impacts, technical feasibility, and economic feasibility; and an overall decision-making strategy.

One major framework, the Lowell Center for Sustainable Production Alternatives Assessment Framework, conceives of alternatives assessment very broadly, as a reflexive problem-solving process that recognizes the social and technical complexity of environmental problems. It emphasizes aspects such as stakeholder participation, transparency of the process, and open discussion of values in decision-making. Most other frameworks are more narrow and primarily focused on technical aspects.

Chemical Alternatives Assessment

Chemical alternatives assessment (or *substitution of hazardous chemicals*) is the use of alternatives assessment for finding safer and environmentally preferable design options to reduce or eliminate the use of hazardous chemical substances. Safer alternatives to hazardous chemicals may simply be other chemical substances, or may involve deeper changes in material or product design. Chemical alternatives assessment aims to provide the basis for well-informed decision-making by thoroughly characterizing chemicals and materials (and other design options) across a wide range of environmental and health impact categories. The rationale for this is to avoid shifting environmental health burdens from one category of impacts to another (e.g., substituting a carcinogenic chemical with a neurotoxic one), and to minimize the unintended consequences of decisions made under conditions of uncertainty and ignorance—in other words, to prevent "regrettable substitutions", where an alternative appeared better based on limited knowledge, but turned out to be worse or equally bad.

An array of methodologies and tools for chemical alternatives assessment have been developed worldwide and have been deployed in a variety of industry sectors. Chemical alternatives assessment frameworks include chemical hazard assessment methods that consider a wide range of hazard endpoints, such as those defined in the Globally Harmonized System.

Examples

- The Massachusetts Toxics Use Reduction Institute has conducted alternatives assessments for the substitution of hazardous substances like lead and formaldehyde.

- The San Francisco Department of Environment has conducted alternatives assessments to identify environmentally and economically preferable alternatives to conventional dry cleaning technology.

- Electronics manufacturer Hewlett-Packard uses the GreenScreen chemical hazard assessment method in product R&D.

Practice of Alternatives Assessment

Scientists from a variety of government agencies, academic institutions, non-profit organizations, and firms have contributed to developing the practice of alternatives assessment.

In the United States

Scientific research in the US federal government has contributed to alternatives assessment frameworks and practices. In 2014 the US National Research Council released a chemical alternatives assessment framework developed by an expert working group. Prior to this, the US Environmental Protection Agency ran a program called *Design for the Environment* (now called Safer Choice), which developed chemical alternatives assessment methodology and created partnerships to undertake numerous alternatives assessments for chemicals of concern in products.

Alternatives assessment has featured in some state-level chemicals policies and regulatory activities. The Massachusetts Toxics Use Reduction Institute has provided public technical assistance for "toxics use reduction planning", which includes alternatives assessment. More recently, the California Department of Toxic Substances Control is implementing new regulations that require firms to undertake alternatives assessments for selected priority chemicals in products. The Interstate Chemicals Clearinghouse, an association of state governments, has also produced its own guide to alternatives assessment.

Some US-based companies have begun to use chemical alternatives assessment to address chemical safety issues in supply chains. For example, firms that participate in the Business-NGO Working Group for Safer Chemicals and Sustainable Materials have produced their own guidance for chemical alternatives assessment.

In the European Union

The Sweden-based non-governmental organization ChemSec has been active in developing resources and tools for the substitution of hazardous chemicals. The Substitution support portal (SUBSPORT) is an EU-based collaboratively-developed resource for chemical substitution. It includes alternatives assessment case studies.

Advanced Oxidation Process

Advanced oxidation processes (abbreviation: AOPs), in a broad sense, are a set of chemical treatment procedures designed to remove organic (and sometimes inorganic) materials in water and waste water by oxidation through reactions with hydroxyl radicals ($\cdot OH$). In real-world applications of wastewater treatment, however, this term usually refers more specifically to a subset of such chemical processes that employ ozone (O_3), hydrogen peroxide (H_2O_2) and/or UV light. One such type of process is called in situ chemical oxidation.

Description

AOPs rely on in-situ production of highly reactive hydroxyl radicals ($\cdot OH$). These reactive species are the strongest oxidants that can be applied in water and can virtually oxidize any compound present in the water matrix, often at a diffusion controlled reaction speed. Consequently, $\cdot OH$ reacts unselectively once formed and contaminants will be quickly and efficiently fragmented and converted into small inorganic molecules. Hydroxyl radicals are produced with the help of one or more primary oxidants (e.g. ozone, hydrogen peroxide, oxygen) and/or energy sources (e.g. ultraviolet light) or catalysts (e.g. titanium dioxide). Precise, pre-programmed dosages, sequences and combinations of these reagents are applied in order to obtain a maximum •OH yield. In general, when applied in properly tuned conditions, AOPs can reduce the concentration of contaminants from several-hundreds ppm to less than 5 ppb and therefore significantly bring COD and TOC down, which earned it the credit of "water treatment processes of the 21st century".

The AOP procedure is particularly useful for cleaning biologically toxic or non-degradable materials such as aromatics, pesticides, petroleum constituents, and volatile organic compounds in waste water. Additionally, AOPs can be used to treat effluent of secondary treated wastewater which is then called tertiary treatment. The contaminant materials are converted to a large extent into stable inorganic compounds such as water, carbon dioxide and salts, i.e. they undergo mineralization. A goal of the waste water purification by means of AOP procedures is the reduction of the chemical contaminants and the toxicity to such an extent that the cleaned waste water may be reintroduced into receiving streams or, at least, into a conventional sewage treatment.

Although oxidation processes involving ·OH have been in use since late 19th century (such as in Fenton reagent, which, however, was an analytical reagent at that time), the utilization of such oxidative species in water treatment did not receive adequate attention until Glaze et al. suggested the possible generation of ·OH "in sufficient quantity to affect water purification" and defined the term "Advanced Oxidation Processes" for the first time in 1987. AOPs still have not been put into commercial use on a large scale (especially in developing countries) even up to today mostly because of the relatively high costs. Nevertheless, its high oxidative capability and efficiency make AOPs a popular technique in tertiary treatment in which the most recalcitrant organic and inorganic contaminants are to be eliminated. The increasing interest in water reuse and more stringent regulations regarding water pollution are currently accelerating the implementation of AOPs at full-scale. There are roughly 500 commercialized AOPs installations around the world at present, mostly in Europe and the United States. Other countries like China are showing increasing interests in AOPs.

Chemical Principles

Generally speaking, chemistry in AOPs could be essentially divided into three parts:

1. Formation of ·OH;

2. Initial attacks on target molecules by ·OH and their breakdown to fragments;

3. Subsequent attacks by ·OH until ultimate mineralization.

The mechanism of ·OH production (Part 1) highly depends on the sort of AOP technique that is used. For example, ozonation, UV/H_2O_2 and photocatalytic oxidation rely on different mechanisms of ·OH generation:

* UV/H_2O_2:

    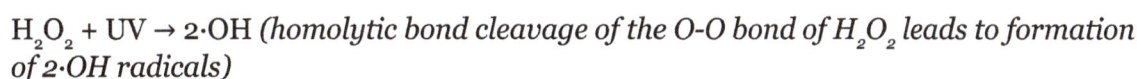

    $H_2O_2 + UV \rightarrow 2\cdot OH$ *(homolytic bond cleavage of the O-O bond of H_2O_2 leads to formation of 2·OH radicals)*

- Ozone based AOP:

$O_3 + HO^- \rightarrow HO_2^- + O_2$ *(reaction between O_3 and a hydroxyl ion leads to the formation of H_2O_2 (in charged form))*

$O_3 + HO_2^- \rightarrow HO_2\cdot + O_3^{-}\cdot$ *(a second O_3 molecule reacts with the HO_2^- to produce the ozonide radical)*

$O_3^{-}\cdot + H^+ \rightarrow HO_3\cdot$ *(this radical gives to ·OH upon protonation)*

$HO_3\cdot \rightarrow \cdot OH + O_2$

the reaction steps presented here are just a part of the reaction sequence.

- Photocatalytic oxidation with TiO_2:

$TiO_2 + UV \rightarrow e^- + h^+$ *(irradiation of the photocatalytic surface leads to an excited electron (e^-) and electron gap (h^+)*

$Ti(IV) + H_2O \rightleftharpoons Ti(IV)\text{-}H_2O$ *(water adsorbs onto the catalyst surface)*

$Ti(IV)\text{-}H_2O + h^+ \rightleftharpoons Ti(IV)\text{-}\cdot OH + H^+$ *the highly reactive electron gap will react with water*

the reaction steps presented here are just a part of the reaction sequence.

Currently there is no consensus on the detailed mechanisms in Part 3, but researchers have cast light on the processes of initial attacks in Part 2. In essence, ·OH is a radical species and should behave like a highly reactive electrophile. Thus two type of initial attacks are supposed to be Hydrogen Abstraction and Addition. The following scheme, adopted from a technical handbook and later refined, describes a possible mechanism of the oxidation of benzene by ·OH.

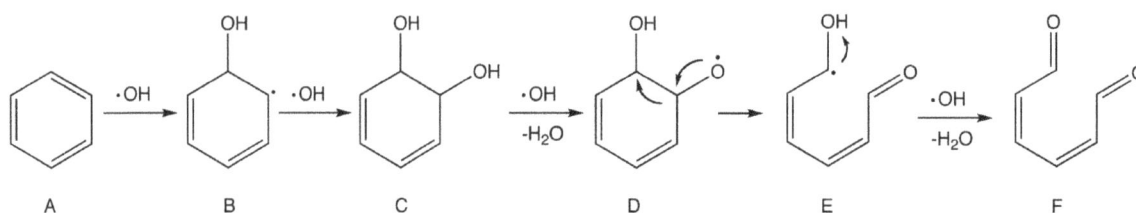

Scheme 1. Proposed mechanism of the oxidation of benzene by hydroxyl radicals

The first and second steps are electrophilic addition that breaks the aromatic ring in benzene (A) and forms two hydroxyl groups (−OH) in intermediate C. Later an ·OH grabs a hydrogen atom in one of the hydroxyl groups, producing a radical species (D) that is prone to undergo rearrangement to form a more stable radical (E). E, on the other hand, is readily attacked by ·OH and eventually forms 2,4-hexadiene-1,6-dione (F). As long as there are sufficient ·OH radicals, subsequent attacks on compound F will continue until the fragments are all converted into small and stable molecules like H_2O and CO_2 in the end, but such processes may still be subject to a myriad of possible and partially unknown mechanisms.

Advantages

AOPs hold several advantages that are unparalleled in the field of water treatment:

- They can effectively eliminate organic compounds in aqueous phase, rather than collecting or transferring pollutants into another phase.

- Due to the remarkable reactivity of $\cdot OH$, it virtually reacts with almost every aqueous pollutant without discriminating. AOPs are therefore applicable in many, if not all, scenarios where many organic contaminants must be removed at the same time.

- Some heavy metals can also be removed in forms of precipitated $M(OH)_x$.

- In some AOPs designs, disinfection can also be achieved, which makes these AOPs an integrated solution to some water quality problems.

- Since the complete reduction product of $\cdot OH$ is H_2O, AOPs theoretically do not introduce any new hazardous substances into the water.

Current Shortcomings

It should be realised that AOPs are not perfect and have several drawbacks.

- Most prominently, the cost of AOPs is fairly high, since a continuous input of expensive chemical reagents is required to maintain the operation of most AOP systems. As a result of their very nature, AOPs require hydroxyl radicals and other reagents proportional to the quantity of contaminants to be removed.

- Some techniques require pre-treatment of wastewater to ensure reliable performance, which could be potentially costly and technically demanding. For instance, presence of bicarbonate ions (HCO_3^-) can appreciably reduce the concentration of $\cdot OH$ due to scavenging processes that yield H_2O and a much less reactive species, $\cdot CO_3^-$. As a result, bicarbonate must be wiped out from the system or AOPs are compromised.

- It is not cost effective to use solely AOPs to handle a large amount of wastewater; instead, AOPs should be deployed in the final stage after primary and secondary treatment have successfully removed a large proportion of contaminants.

Future

Since AOPs were first defined in 1987, the field has witnessed a rapid development both in theory and in application. So far, TiO_2/UV systems, H_2O_2/UV systems, and Fenton, photo-Fenton and Electro-Fenton systems have received extensive scrutiny. However, there are still many research needs on these existing AOPs.

Recent trends are the development of new, modified AOPs that are efficient and economical. In fact, there has been some studies that offer constructive solutions. For instance, doping TiO_2 with non-metallic elements could possibly enhance the photocatalytic activity; and implementation of ultrasonic treatment could promote the production of hydroxyl radicals.

California Green Chemistry Initiative

The California Green Chemistry Initiative (CGCI) is a six-part initiative to reduce public and environmental exposure to toxins through improved knowledge and regulation of chemicals; two parts became statute in 2008. The other four parts were not passed, but are still on the agenda of the California Department of Toxic Substances Control green ribbon science panel discussions. The two parts of the California Green Chemistry Initiative that were passed are known as AB 1879 (Chapter 559, Statutes of 2008): Hazardous Materials and Toxic Substances Evaluation and Regulation and SB 509 (Chapter 560, Statutes of 2008): Toxic Information Clearinghouse. Implementation of CGCI has been delayed indefinitely beyond the January 1, 2011.

Purpose

Green chemistry is the design of chemical products and processes that reduce or eliminate the use and generation of hazardous substances. Green chemistry is based upon twelve principles, identified in "Green Chemistry: Theory and Practice" and adopted by the US Environmental Protection Agency (EPA). It is an innovative technology which encourages the design of safer chemicals and products and minimizes the impact of wastes through increased energy efficiency, the design of chemical products that degrade after use and the use of renewable resources (instead of non-renewable fossil fuel such as petroleum, gas and coal). The Office of Pollution Prevention and Toxics (OPPT), created under the United States Pollution Prevention Act of 1990, promotes the use of chemistry for pollution prevention through voluntary, non-regulatory ' partnerships with academia, industry, other government agencies, and non-governmental organizations. The United States Environmental Protection Agency (EPA) promotes green chemistry as overseen by the OPPT. The California Green Chemistry Initiative moves beyond voluntary partnerships and voluntary information disclosure to require industry reporting and public disclosure.

Overview

The United States Environmental Protection Agency's most important law to regulate the production, use and disposal of chemicals is the Toxic Substances Control Act of 1976 (TSCA). Over the years, TSCA has fallen behind the industry it is supposed to regulate and is an inadequate tool for providing the protection against today's chemical risks. Green chemistry represents a major paradigm shift in industrial manufacturing as it is a proactive "cradle-to-cradle" approach that focuses environmental protection at the design stage of production processes.

In 2008, California governor Arnold Schwarzenegger signed two joined bills, AB 1879 and SB 507, which created California's Green Chemistry Initiative (CGCI). AB 1879 increases regulatory authority over chemicals in consumer products. The law established an advisory panel of scientists, known as the green ribbon science panel, to guide research in chemical policy, create regulations for assessing alternatives, and set up an internet database of research on toxins. SB 509 was designed to ensure that information regarding the hazard traits, toxicological and environmental endpoints, and other vital data is available to the public, to businesses, and to regulators in a

Toxics Information Clearinghouse. This legislation marks the biggest leap forward in California chemicals policy in nearly two decades and is intended to improve the health and safety of all Californians by providing the Department of Toxic Substances Control (DTSC) with the authority to control toxic substances in consumer products.

The bills were scheduled to go into regulatory affect January 1, 2011 with the adoption of the Green Chemistry Initiative. California has postponed the initiative, indefinitely, due to concerns raised by stakeholders and more specifically, controversial last minute changes in the final draft. The final or third draft contains substantial revisions, including scaled back manufacturer and retailer compliance requirements that were not well received by the environmental community. Assemblyman Mike Feur and several authors of AB 1879, assert that last minute changes by the California DTSC have drastically weakened the Green Chemistry Initiative and limited its scope. They are most concerned with the change to require the state to prove that a chemical is harmful before being regulated, mirroring what is currently required at the Federal level by TSCA. The original draft advocated a precautionary principle, or "cradle-to-cradle" approach. Environmentalists fear that CGCI will not remove chemicals off the shelves, but instead will create "paralysis by analysis" as companies litigate against the DTSC over unfavorable decisions.

Physical and Social Causes

Traditional Methods of Dealing with Wastes

Society historically managed its industrial and municipal wastes by disposal or incineration. Chemical regulation occurs only after a product is identified as hazardous. This problem-specific approach has led to the release of thousands of potentially harmful chemicals in our environment. Chemical regulation is a continuous game of catch up, in which banned chemicals are replaced with new chemicals that may be just as or more toxic. Many environmental laws are still based on the industrial production model of cradle-to-grave. The term "cradle-to-grave" is used to describe and assess the life-cycle of products, from raw material extraction through materials processing, manufacture, distribution, use and disposal. This traditional approach to chemicals management has serious environmental drawbacks because it does not consider what happens to a product after it is disposed of. The Resource Conservation and Recovery Act (RCRA) of 1976, exemplifies a cradle-to-grave management approach of hazardous waste. RCRA has been largely ineffective because its emphasis is on dealing with waste after it has been created; meanwhile emphasis on waste reduction is minimal. Waste does not disappear, it is simply transported elsewhere. Costly and burdensome hazardous waste disposal in the US has encouraged the exportation of hazardous waste to poor counties and developing nations willing to accept the waste for a fee.

The Green Chemistry initiative instead employs a cradle-to-cradle approach, representing a major paradigm shift in environmental policy and provides a proactive solution to toxic waste. The Earth's capacity to accept toxic waste is practically nonexistent. The disposal of hazardous wastes is not the root problem but rather, the root symptom. The critical issue is the creation of toxic wastes. Requiring manufacturers to consider chemical exposure during manufacturing, throughout product use and after disposal, encourages the production of safer products.

Consumption and Wastes

By the time we find a product on a market shelf, 90% of the resources used to create that product was regarded as waste. This accounts for about 136 pounds of resources a week consumed by the average American and 2,000 pounds of waste support that consumption. As the population grows and the economy expands more and more products will be created, consumed, and disposed. Many negative externalities are related to the environmental consequences of production and use, including air pollution, anthropogenic climate change and water pollution. Under the current cycle of production, toxic chemical byproducts will continue to be produced and unleashed on our environment. It is important to carefully consider how toxic wastes are created in order to forgo the possibility of a world that is unsuitable for human life.

Transparency Issues

One of the biggest failures in market transactions is the imbalance of information that is provided to consumer via producer. "Information asymmetry" is an economic concept that is used to explain this failure: it deals with the study of decisions in transactions where one party has more or better information than the other. Due to a lack of information transparency, the public may lack vital information about the health and safety of products found on supermarket shelves. This lack of information may have led to a reversed purchasing decision. Yet without such labeling, consumers must make assumptions based on things like price or expertise. For example, one apple juice brand may be assumed healthier because it cost more and because the brand is advertised as "healthy" and "recommended by mothers". Further, it may be assumed that the product is safe for consumption if it is sitting on a grocery store shelf and probably would not be approved by the government if it contained harmful chemicals. Assumptions such as these could inform a typical purchasing decision, despite their inaccuracy. Perhaps given more information, the same brand of apple juice would be less desirable if information on unhealthy preservatives, additives or pesticide residues was easily obtained. To make market transactions more efficient, the government could force more accurate labeling about products, laws could require companies to be more transparent, and the government could require that advertising be less persuasive and more informative. The Green Chemistry Initiative of California would address transparency issues by creating a public chemical inventory and requiring more stringent regulation of chemicals that may be toxic. The CGCI Draft Report suggests a green labeling system to identify consumer products with ingredients harmful to human health and the environment.

Stakeholder Involvement

The United States is the world leader in chemicals manufacturing. As a multibillion-dollar industry, the chemical industry has a leading role in the US economy and because of this, a high level of influence in federal decision-making. Central to the modern world economy, it converts raw materials (oil, natural gas, air, water, metals, and minerals) into more than 70,000 different products. The chemical industry—producers of chemicals, household cleansers, plastics, rubber, paints and explosives, keeps a watchful eye on issues including environmental and health policy, taxes and trade. The industry is often the target of environmental groups, which charge that chemicals and chemical waste are polluting the air and water supply. And like most industries with pollution problems, chemical manufacturers oppose meddlesome government regulations that make it more difficult and expensive for them to do business. So do most Republicans, which is why this industry gives nearly three-fourths of its campaign contributions to the GOP. In addition to campaign contributions to elected officials and candidates, companies, labor unions, and other organizations spend billions of dollars each year to lobby Congress and federal agencies. Some special interests retain lobbying firms, many of them located along Washington's legendary K Street; others have lobbyists working in-house.

According to website *Opensecrets,* the total number of clients lobbying for the chemical industry in 2010 was 143, which is the highest number in history. The first group on this list, American Chemistry Council spent $8,130,000 lobbying last year and Crop America, which comes second, spent $2,291,859 lobbying last year, FMC Corporation spent $1,230,000 and Koch Industries spent $8,070,000. The Chemical Industry wants limited testing of chemicals, more lengthy and costly studies of chemicals already proven to be dangerous, and an assumption that we are only exposed to one chemical at a time, and from one source at a time.

According to *Safer Chemicals, Healthy Families,* a broad coalition of groups, including major environmental organizations like the Natural Resources Defense Council and the Environmental Defense Fund, health organizations like the Learning Disabilities Association, Breast Cancer Fund, and the Autism Society of America, health professionals and providers like the American Nurses Association, Planned Parenthood Federation of America, and the Mt. Sinai Children's Environmental Health Center, and concerned parents groups like MomsRising: there is growing national momentum and pressure to change the Toxic Substances Control Act (TSCA), our federal system for overseeing chemical safety, which has not been updated in thirty-five years. Polling data indicates overwhelming support for chemical regulation nationwide. According to polling data conducted by the Mellman Group, 84% say that "tightening controls" on chemical regulation is important, with 50% of those calling it "very important." Public Health Advocates want public disclosure of safety information for all chemicals in use, prompt action to phase out or reduce the most dangerous chemicals, deciding safety based on real world exposure to all sources of toxic chemicals.

History

In 2008, California Governor Arnold Schwarzenegger signed two state bills authorizing the state to identify toxic chemicals in industry and consumer products and analyze alternatives. AB 1879, written by Assemblyman Mike Feur, a Los Angeles Democrat, requires the state Department of Toxic Substances Control to assess chemicals and prioritize the most toxic for possible restrictions or bans. The environmental policy council, made up of heads of all state environmental protection agency boards and departments will oversee the program. SB 509, by Senator Joe Simitian, a Palo Alto Democrat, creates an online toxics information clearinghouse with information about the hazards of thousands of chemicals used in California. These bills are intended to put an end to chemical-by-chemical bans and remove harmful products at the design stage. The regulations are expected to motivate manufacturers of consumer products containing chemicals of concern to seek safer alternatives.

Supporters of the bill include the California Association of Professional Scientists, the Chemical Industry Council of California, DuPont, BIOCOM, Grocery Manufacturers Association, the Breast Cancer Fund, Catholic Healthcare West, in addition to a broad array of environmental groups such as the Coalition for Clean Air, the Environmental Defense Fund, the Natural Resources Defense

Council. The American Electronics Association (AEA) and Ford spoke in opposition to the bill, each requesting an exemption from its provisions. Also opposing were environmental justice advocates who indicated the bill did not go far enough. Meanwhile, large trade associations such as Consumer Specialty Products Association, Western States Petroleum Association, American Chemistry Council, CA Manufacturers and Technology Association, and CA Chamber of Commerce officially withdrew opposition to the measures.

Due to outdated and inefficient or otherwise voluntary chemical regulation at the Federal level, the State of California has decided to take regulation into its own hands and develop stricter, environmentally-informed methodologies for dealing with the production of toxic wastes. California's economy is the largest of any state in the US, and is the eighth largest economy in the world. This position gives California an advantage when it comes to environmental standards: the impact of chemical regulation statewide can have a broader impact nationwide if manufacturers desire to stay competitive in California's market. The Green Chemistry Initiative forces statewide industries to comply with greener standards of production, which may spark innovation on a wider basis.

The Green Chemistry initiative aims to regulate the creation and use of materials hazardous to human health and the environment by encouraging innovative design and manufacturing, and ultimately safer consumer product alternatives. To develop the regulatory framework, DTSC held a number of stakeholder and public workshops and invited direct public participation in the drafting of regulations on a wiki website. DTSC reportedly received over 57,000 comments and over 800 regulatory suggestions. Regulatory suggestions included industry assessments of risk and safety, alternative chemicals and life-cycle assessments and mandatory industry reporting, full public disclosure of substances contained in products, a green labelling program that would inform consumers of the potential health and environmental impacts of the chemicals contained in products and a mandated surcharge on chemicals and products to support a fund to address environmental problems. In December 2008, DTSC announced six policy recommendations for the Green Chemistry Initiative. In brief, those recommendations are:

1. expand pollution prevention

2. develop green chemistry workforce education and training, research and development, technology transfer

3. online product ingredient network

4. online toxics clearing house

5. accelerate the quest for safer products

6. move toward cradle to cradle economy

Two of the six recommendations from this report were adopted: AB 1879 requires the DTSC to implement regulations to identify and prioritize chemicals of concern, evaluate alternatives, and specify regulatory responses where chemicals are found in products. SB 509 requires an online, public toxics information clearinghouse that includes science-based information on the toxicity and hazard traits of chemicals used in daily life. Essentially the recommended policy methods include authority tools that would regulate the approval on new chemicals in a more cautious manner as well as mandate the decimation of information, as provided by manufacturers to the public; innovation would be encouraged under this paradigm to replace harmful chemicals with greener alternatives and the California government would fund programs to help industries produce greener chemicals. Secondly, capacity or learning tools would be provided to the public in the form of the online database, giving the tools so that they have better ability to make market decisions that reflect their interests.

Criticism

Environmentalists say the amended regulations won't remove toxic products from the shelves and will create "paralysis by analysis," as industries can litigate against DTSC over unfavorable department decisions. Activists say California was poised to lead the way on toxics regulation but now is faced with potentially one of the weakest chemical-regulatory mechanisms in the nation. According to CHANGE (Californians for a Healthy & Green Economy), the revised regulation is a betrayal of the Green Chemistry promise and ignores two years of public input, while caving to backroom industry lobbying. Furthermore, it is a betrayal to public interest groups, businesses, and residents of California and legislators who supported the intent of this bill, to protect Californians and spur a healthy, innovative green economy. Environmentalists say the toxics department gutted the initiative at the behest of the chemical industry, and then put out the changes for public comment during a 15-day period just before Thanksgiving. This was a violation of the law requiring a 45-day public comment period when a substantial reworking of state regulations is proposed. The new Director of California's Department of Toxic Substance Control, Debbie Raphael, announced that mid-October 2011 is the new target date for new draft regulations to implement California's Green Chemistry Law and new draft guidelines were issued October 31, 2011. The public comment period for the latest version of the draft regulations ends December 30, 2011.

Implementation of CGCI has been delayed indefinitely beyond the January 1, 2011 deadline due to issues that arose after public review of the third draft. The third draft, which was made public December 2010, contains substantial revisions, including scaled back manufacturer and retailer compliance requirements that were not well received by the environmental community. DTSCs newest draft has made the following changes:

- All references of nanotechnology are excluded (nano referring to materials with dimensions of 1,000 nanometers or smaller); this change is significant because it would have been considered the most significant attempt to regulate nanomaterials based on environmental or health impacts.

- The new draft redefines "responsible entities," which originally referred to the entire business chain of consumer products distribution, including manufacturers, brand name owners, importers, distributors, and retailers, "responsible entities is now limited to manufacturers and retailers .

- DTSC prioritizes Children's products, personal care products and household products until 2016, after that point all consumer products.

- The new proposed regulations also eliminate the requirement that the DTSC develop a list of chemicals of consideration and products under consideration.

- New timeline for implementation of regulations

International Conference on Green Chemistry

The International IUPAC Conferences on Green Chemistry (ICGCs) gather several hundreds scientists, technologists, and experts from all over the world with the aim to exchange and disseminate new ideas, discoveries, and projects on green chemistry and a sustainable development. After mid twentieth century, an increasingly general consensus acknowledges that these subjects play a unique role in mapping the way ahead for the humankind progress. Typical topics discussed in these IUPAC Conferences are:

- bio-based renewable chemical resources, bio-inspired materials and nanomaterials, bio-based polymers;

- polymer composites and natural surfactants;

- green solvents, catalysts, and synthetic methodologies (e.g., microwaves, ultrasounds, solid state synthesis), biocatalysis and biotransformations;

- biofuels and chemistry for improved energy harvesting;

- materials for sustainable construction and cultural heritage;

- pollution prevention;

- metrics, evaluation, education, and communication of green chemistry.

History

In 2006 the International Union of Pure and Applied Chemistry (IUPAC) promoted the organization of the 1st International IUPAC Conference on Green-Sustainable Chemistry (ICGC-1). This conference, started in collaboration with the German Chemical Society (GDCh), was a major acknowledgement by IUPAC of the relevance of green chemistry. The Special Topic Issue on Green Chemistry in *Pure and Applied Chemistry* and the starting of a Subcommittee on Green Chemistry, operating in the IUPAC Division of Organic and Biomolecular Chemistry, were two important landmarks towards that acknowledgement. ICGC-1 registered the presence of over 450 participants from 42 countries and proceedings were published in *Pure and Applied Chemistry*. This Conference then became a biannual appointment that continuously attracted several hundreds scientists and technologists from academia, research institutes, and industries.

List of Symposia

N.	Year	City	Country	Date	Chair(s)	Proceedings published in (Event announced by)
1	2006	Dresden	Germany	10–15 September	Piero Tundo and Wolfgang Hoelderich	*Pure Appl. Chem.* 2007, 79(11), 1833–2105 (*Chem. Int.* 2005, 27(6))

2	2008	Moscow	Russia	14–20 September	Valery V. Lunin and Ekaterina Lokteva	*Pure Appl. Chem.* 2009, 81(11), 1961-2129 (*Chem. Int.* 2007, 29(6))
3	2010	Ottawa	Canada	15–19 August	Philip Jessop	*Pure Appl. Chem.* 2011, 83(7), 1343-1406 (*Chem. Int.* 2009, 31(4))
4	2012	Sao Carlos	Brazil	25–29 August	Arlene Correa and Vania Zuin	*Pure Appl. Chem.* 2013, 85(8), 1611–1710 (*Chem. Int.* 2011, 33(2))
5	2014	Durban	South Africa	17–21 August	Liliana Mammino	*Pure Appl. Chem.* to be published on 2016
6	2016	Venice	Italy	4–8 September	Piero Tundo	*Pure Appl. Chem.* to be published on 2018

Vinyloop

Vinyloop® is a physical plastic recycling process for polyvinyl chloride (PVC). It is based on dissolution in order to separate PVC from other materials or impurities.

Background

A major factor of the recycling of polyvinyl chloride waste is the purity of the recycled material. In most composite materials, PVC is among several other materials, such as wood, metal, or textile. To make new products from the recycled PVC, it is necessary to separate it from other materials. Traditional recycling methods are not sufficient and expensive because this separation has to be done manually and product by product.

Vinyloop is a recycling process which separates PVC from other materials through a process of dissolution, filtration and separation of contamination. A solvent is used in a closed loop to elute PVC from the waste. This makes it possible to recycle composite structure PVC waste, which would normally be incinerated or put in a landfill site.

Process

The process consists of the following steps:

1. Pretreatment: waste plastics are cleaned, ground and mixed

2. Dissolution: a specific solvent is used to selectively dissolve the PVC compound in a closed loop

3. Filtration: impurities which have not been are removed through filtration can be dissolved – they are separated by type of material by filtration, centrifugation and decantation. After separation, the secondary materials are washed with pure solvent to dissolve all remaining PVC compounds

4. Precipitation of the regenerated PVC compound: the solution of PVC is recovered in a precipitation tank, where steam is injected to evaporate the solvent and precipitate the PVC. The PVC compound is separated in the form of aqueous effluent. and dried.

5. Drying: after recovering the excess water from the slurry, the wet PVC goes to a dryer.

Possible products made from recycled PVC are coatings for waterproofing membranes, pond foils, shoe soles, hoses, diaphragms tunnel, coated fabrics, and PVC sheets. It is an attempt to solve the recycling waste problemacy of PVC products.

Ecological Importance

"The environmental performance of PVC recycling (Vinyloop) is a lot better than new production in most of the impact categories." Vinyloop-based recycled PVC's primary energy demand is 46 percent lower than conventional produced PVC. The global warming potential is 39 percent lower.

The Vinyloop process has been selected to recycle membranes of different temporary venues of the London Olympics 2012. Roofing covers of the Olympic Stadium, the Water Polo Arena, the London Aquatics Centre and the Royal Artillery Barracks will be deconstructed and a part will be recycled in the Vinyloop process.

As we have done in the past with materials such as timber and concrete, we want to use the opportunity of hosting the London 2012 Games to work with industry to set new standards. In this case this may help move the industry towards more sustainable manufacture, use and disposal of PVC fabrics.

Dan Epstein Head of Sustainable Development at Olympic Delivery Authority (ODA)

References

- O'Brien, Mary (2000). Making better environmental decisions: An alternative to risk assessment. Cambridge: MIT Press. ISBN 0-262-15051-4.

- Geiser, Ken (2015). Chemicals without harm: Policies for a sustainable world. Urban and industrial environments. Cambridge, Mass.: The MIT Press. ISBN 978-0-262-51206-0.

- National Research Council (US) (2014). A framework to guide selection of chemical alternatives. Washington, D.C: The National Academies Press. ISBN 978-0-309-31013-0.

- Beltrán, Fernando J. (2004). Ozone Reaction Kinetics for Water and Wastewater Systems. CRC Press, Florida. ISBN 1-56670-629-7.

- Business-NGO Working Group for Safer Chemicals and Sustainable Materials (BizNGO) (2011-11-30), BizNGO chemical alternatives assessment protocol

4

Interdisciplinary Aspects of Green Chemistry

The interdisciplinary aspects of green chemistry are chemical synthesis, process chemistry and biochemistry. Chemical synthesis is the inducement of a chemical reaction and this execution is done in order to obtain a product. In today's usage, this simply means that the process is reproducible and can be worked with in multiple laboratories. The major aspects of green chemistry are discussed in the following chapter.

Chemical Synthesis

Chemical synthesis is a purposeful execution of chemical reactions to obtain a product, or several products. This happens by physical and chemical manipulations usually involving one or more reactions. In modern laboratory usage, this tends to imply that the process is reproducible, reliable, and established to work in multiple laboratories.

Introduction

A chemical synthesis—*synthesis,* in its present meaning originating with chemist Hermann Kolbe—begins with the careful selection of the target, based on the relationship of *possible* targets of chemical, functional, or therapeutic interest to the broader aims of the research effort, whether industrial or academic; various further practical concerns come into play, including manpower available for the campaign, as well as availability of material resources (necessary equipment, starting materials and bulk chemicals, etc.), and along with these, budget for these practical necessities. For instance, a prior developed reaction methodology may highlight particular man-made or natural compounds that would serve the purposes of the effort, in highlighting the breadth of the new methodology; alternatively, grant or unit budgetary or timeline constraints may necessitate aiming for a simpler member of a family of complex natural products, or practical constraints such as the lack of suitable starting materials may necessitate a semi-synthesis over a total synthesis, etc. Once the target or targets are established, the next critical phase begins, that of *synthetic design*, typically in modern efforts, in the area of organic synthesis, using retrosynthetic analysis, as championed by E.J. Corey and others.

An eventual step is selection of compounds that are known as reagents or reactants. Various reaction types can be applied to these to synthesize the product, or an intermediate product. This requires mixing the compounds in a reaction vessel such as a chemical reactor or a simple round-bottom flask. Many reactions require some form of work-up procedure before the final product is isolated. The isolation (purification) of the product then proceeds via a variety of methods.

The amount of product in a chemical synthesis is the reaction yield. Typically, chemical yields are expressed as a weight in grams (in a laboratory setting) or as a percentage of the total theoretical quantity of product that could be produced. A side reaction is an unwanted chemical reaction taking place that diminishes the yield of the desired product.

Strategies

Many strategies exist in chemical synthesis that go beyond converting reactant A to reaction product B in a single step. In multistep synthesis, a chemical compound is synthesised though a series of individual chemical reactions, each with their own work-up. For example, a laboratory synthesis of paracetamol can consist of three individual synthetic steps. In cascade reactions multiple chemical transformations take place within a single reactant, in multi-component reactions up to 11 different reactants form a single reaction product and in a telescopic synthesis one reactant goes through multiple transformations without isolation of intermediates.

Organic Synthesis

Organic synthesis is a special branch of chemical synthesis dealing with the synthesis of organic compounds. In the total synthesis of a complex product it may take multiple steps to synthesize the product of interest, and inordinate amounts of time. Skill in organic synthesis is prized among chemists and the synthesis of exceptionally valuable or difficult compounds has won chemists such as Robert Burns Woodward the Nobel Prize for Chemistry. If a chemical synthesis starts from basic laboratory compounds and yields something new, it is a purely synthetic process. If it starts from a product isolated from plants or animals and then proceeds to new compounds, the synthesis is described as a semisynthetic process.

Process Chemistry

Process chemistry is the arm of pharmaceutical chemistry concerned with the development and optimization of a synthetic scheme and pilot plant procedure to manufacture compounds for the drug development phase. Process chemistry is distinguished from medicinal chemistry, which is the arm of pharmaceutical chemistry tasked with designing and synthesizing molecules on small scale in the early drug discovery phase.

Medicinal chemists are largely concerned with synthesizing a large number of compounds as quickly as possible from easily tunable chemical building blocks (usually for SAR studies). In general, the repertoire of reactions utilized in discovery chemistry is somewhat narrow (for example, the Buchwald-Hartwig amination, Suzuki coupling and reductive amination are commonplace reactions). In contrast, process chemists are tasked with identifying a chemical process that is safe, cost and labor efficient, "green," and reproducible, among other considerations. Oftentimes, in searching for the shortest, most efficient synthetic route, process chemists must devise creative synthetic solutions that eliminate costly functional group manipulations and oxidation/reduction steps.

This section focuses exclusively on the chemical and manufacturing processes associated with the production of small molecule drugs. Biological medical products (more commonly called "biologics") represent a growing proportion of approved therapies, but the manufacturing processes of these products are beyond the scope of this article. Additionally, the many complex factors associated with chemical plant engineering (for example, heat transfer and reactor design) and drug formulation will be treated cursorily.

Process Chemistry Considerations

Cost efficiency is of paramount importance in process chemistry and, consequently, is a focus in the consideration of pilot plant synthetic routes. The drug substance that is manufactured, prior to formulation, is commonly referred to as the active pharmaceutical ingredient (API) and will be referred to as such herein. API production cost can be broken into two components: the "material cost" and the "conversion cost." The ecological and environmental impact of a synthetic process should also be evaluated by an appropriate metric (e.g. the EcoScale).

An ideal process chemical route will score well in each of these metrics, but inevitably tradeoffs are to be expected. Most large pharmaceutical process chemistry and manufacturing divisions have devised weighted quantitative schemes to measure the overall attractiveness of a given synthetic route over another. As cost is a major driver, material cost and volume-time output are typically weighted heavily.

Material Cost

The material cost of a chemical process is the sum of the costs of all raw materials, intermediates, reagents, solvents and catalysts procured from external vendors. Material costs may influence the selection of one synthetic route over another or the decision to outsource production of an intermediate.

Conversion Cost

The conversion cost of a chemical process is a factor of that procedure's overall efficiency, both in materials and time, and its reproducibility. The efficiency of a chemical process can be quantified by its atom economy, yield, volume-time output, and environmental factor (E-factor), and its reproducibility can be evaluated by the Quality Service Level (QSL) and Process Excellence Index (PEI) metrics.

An illustrative example of atom economy using the Claisen rearrangement and Wittig reaction.

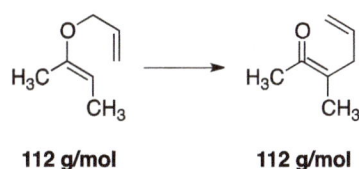

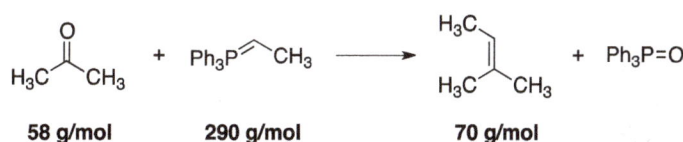

Atom Economy

The atom economy of a reaction is defined as the number of atoms from the starting materials that are incorporated into the final product. Atom economy can be viewed as an indicator of the "efficiency" of a given synthetic route.

$$AE = \frac{MW(\text{product})}{\sum MW(\text{raw materials})} \times 100\%$$

For example, the Claisen rearrangement and the Diels-Alder cycloaddition are examples of reaction that are 100 percent atom economical. On the other hand, a prototypical Wittig reaction has especially poor atom economy.

Process synthetic routes should be designed such that atom economy is maximized for the entire synthetic scheme. Consequently, "costly" reagents such as protecting groups and high molecular weight leaving groups should be avoided where possible. An atom economy value in the range of 70 to 90 percent for an API synthesis is ideal, but it may be impractical or impossible to access certain complex targets within this range. Nevertheless, atom economy is a good metric to compare two routes to the same molecule.

Yield

Yield is defined as the amount of product obtained in a chemical reaction the yield of practical significance in process chemistry is the isolated yield—the yield of the isolated product after all purification steps. In a final API synthesis, isolated yields of 80 percent or above for each synthetic step are expected. The definition of an acceptable yield depends entirely on the importance of the product and the ways in which available technologies come together to allow their efficient application; yields approaching 100% are termed quantitative, and yields above 90% are broadly understood as excellent.

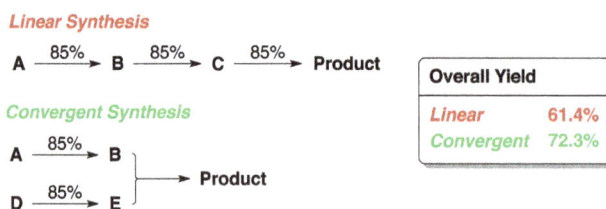

An illustrative example of convergent synthesis.

There are several strategies that are employed in the design of a process route to ensure adequate overall yield of the pharmaceutical product. The first is the concept of convergent synthesis. Assuming a very good to excellent yield in each synthetic step, the overall yield of a multistep reaction can be maximized by combining several key intermediates at a late stage that are prepared independently from each other.

Another strategy to maximize isolated yield (as well as time efficiency) is the concept of telescoping synthesis (also called one-pot synthesis). This approach describes the process of eliminating work-up and purification steps from a reaction sequence, typically by simply adding reagents sequentially to a reactor. In this way, unnecessary losses from these steps can be avoided.

Finally, to minimize overall cost, synthetic steps involving expensive reagents, solvents or catalysts should be designed into the process route as late stage as possible, to minimize the amount of reagent used.

In a pilot plant or manufacturing plant setting, yield can have a profound effect on the material cost of an API synthesis, so the careful planning of a robust route and the fine-tuning of reaction conditions are crucially important. After a synthetic route has been selected, process chemists will subject each step to exhaustive optimization in order to maximize overall yield. Low yields are typically indicative of unwanted side product formation, which can raise red flags in the regulatory process as well as pose challenges for reactor cleaning operations.

Volume-time Output

The volume-time output (VTO) of a chemical process represents the cost of occupancy of a chemical reactor for a particular process or API synthesis. For example, a high VTO indicates that a particular synthetic step is costly in terms of "reactor hours" used for a given output. Mathematically, the VTO for a particular process is calculated by the total volume of all reactors (m³) that are occupied times the hours per batch divided by the output for that batch of API or intermediate (measured in kg).

$$VTO = \frac{\text{nominal volume of all reactors}[m^3] * \text{time per batch}[h]}{\text{output per step}[kg]}$$

The process chemistry group at Boehringer-Ingelheim, for example, targets a VTO of less than 1 for any given synthetic step or chemical process.

Additionally, the raw conversion cost of an API synthesis (in dollars per batch) can be calculated from the VTO, given the operating cost and usable capacity of a particular reactor. Oftentimes, for large-volume APIs, it is economical to build a dedicated production plant rather than to use space in general pilot plants or manufacturing plants.

Environmental Factor (e-factor) and Process Mass Intensity (PMI)

Both of these measures, which capture the environmental impact of a synthetic reaction, intend to capture the significant and rising cost of waste disposal in the manufacturing process. The E-factor for an entire API process is computed by the ratio of the total mass of waste generated in the synthetic scheme to the mass of product isolated.

$$E = \frac{\sum \text{mass of waste}}{\text{mass of isolated product}} = \frac{\sum \text{mass of materials} - \text{mass of isolated product}}{\text{mass of isolated product}}$$

A similar measure, the process mass intensity (PMI) calculates the ratio of the total mass of materials to the mass of the isolated product.

$$PMI = \frac{\sum \text{mass of materials}}{\text{mass of isolated product}} = E + 1$$

For both metrics, all materials used in all synthetic steps, including reaction and workup solvents, reagents and catalysts, are counted, even if solvents or catalysts are recycled in practice. Inconsistencies in E-factor or PMI computations may arise when choosing to consider the waste associated with the synthesis of outsourced intermediates or common reagents. Additionally, the environmental impact of the generated waste is ignored in this calculation; therefore, the environmental quotient (EQ) metric was devised, which multiplies the E-factor by an "unfriendliness quotient" associated with various waste streams. A reasonable target for the E-factor or PMI of a single synthetic step is any value between 10 and 40.

Quality Service Level (QSL)

The final two "conversion cost" considerations involve the reproducibility of a given reaction or API synthesis route. The quality service level (QSL) is a measure of the reproducibility of the quality of the isolated intermediate or final API. While the details of computing this value are slightly nuanced and unimportant for the purposes of this article, in essence, the calculation involves the ratio of satisfactory quality batches to the total number of batches. A reasonable QSL target is 98 to 100 percent.

Process Excellence Index (PEI)

Like the QSL, the process excellence index (PEI) is a measure of process reproducibility. Here, however, the robustness of the procedure is evaluated in terms of yield and cycle time of various operations. The PEI yield is defined as follows:

$$\text{PEI yield} = \frac{\text{average yield} * 100\%}{\text{aspiration level yield}} = \frac{\text{average yield} * 100\%}{\dfrac{\text{median yield} + \text{best yield}}{2}}$$

In practice, if a process is high-yielding and has a narrow distribution of yield outcomes, then the PEI should be very high. Processes that are not easily reproducible may have a higher aspiration level yield and a lower average yield, lowering the PEI yield.

Similarly, a PEI cycle time may be defined as follows:

$$\text{PEI cycle time} = \frac{\text{aspiration level cycle time} * 100\%}{\text{average cycle time}} = \frac{\dfrac{\text{median cycle time} + \text{best cycle time}}{2}}{\text{average cycle time}}$$

For this expression, the terms are inverted to reflect the desirability of shorter cycle times (as opposed to higher yields). The reproducibility of cycle times for critical processes such as reaction, centrifugation or drying may be critical if these operations are rate-limiting in the manufacturing plant setting. For example, if an isolation step is particularly difficult or slow, it could become the bottleneck for an API synthesis, in which case the reproducibility and optimization of that operation become critical.

For an API manufacturing process, all PEI metrics (yield and cycle times) should be targeted at 98 to 100 percent.

EcoScale

In 2006, Van Aken, et al. developed a quantitative framework to evaluate the safety and ecological impact of a chemical process, as well as minor weighting of practical and economical considerations. Others have modified this EcoScale by adding, subtracting and adjusting the weighting of various metrics. Among other factors, the EcoScale takes into account the toxicity, flammability and explosive stability of reagents used, any nonstandard or potentially hazardous reaction conditions (for example, elevated pressure or inert atmosphere), and reaction temperature. Some EcoScale criteria are redundant with previously considered criteria (e.g. E-factor).

Synthetic Case Studies

Boehringer Ingelheim HCV Protease Inhibitor (BI 201302)

Macrocyclization is a recurrent challenge for process chemists, and large pharmaceutical companies have necessarily developed creative strategies to overcome these inherent limitations. An interesting case study in this area involves the development of novel NS3 protease inhibitors to treat Hepatitis C patients by scientists at Boehringer-Ingelheim. The process chemistry team at BI was tasked with developing a cheaper and more efficient route to the active NS3 inhibitor BI 201302, a close analog of BILN 2061. Two significant shortcomings were immediately identified with the initial scale-up route to BILN 2061, depicted in the scheme below. The macrocyclization step posed four challenges inherent to the cross-metathesis reaction.

1. High dilution is typically necessary to prevent unwanted dimerization and oligomerization of the diene starting material. In a pilot plant setting, however, a high dilution factor translates into lower throughput, higher solvent costs and higher waste costs.

2. High catalyst loading was found to be necessary to drive the RCM reaction to completion. Because of high licencing costs of the ruthenium catalyst that was used (1st generation Hoveyda catalyst), a high catalyst loading was financially prohibitive. Recycling of the catalyst was explored, but proved impractical.

3. Long reaction times were necessary for reaction completion, due to slow kinetics of the reaction using the selected catalyst. It was hypothesized that this limitation could be overcome using a more active catalyst. However, while the second-generation Hoveyda and Grubbs catalysts were kinetically more active than the first-generation catalyst, reactions using these catalysts formed large amounts of dimeric and oligomeric products.

4. An epimerization risk under the cross-metathesis reaction conditions. The process chemistry group at Boehringer-Ingelheim performed extensive mechanistic studies showing that epimerization most likely occurs through a ruthenacyclopentene intermediate. Furthermore, the Hoveyda catalyst employed in this scheme minimizes epimerization risk compared with the analogous Grubbs catalyst.

Additionally, the final double S_N2 sequence to install the quinoline heterocycle was identified as a secondary inefficiency in the synthetic route.

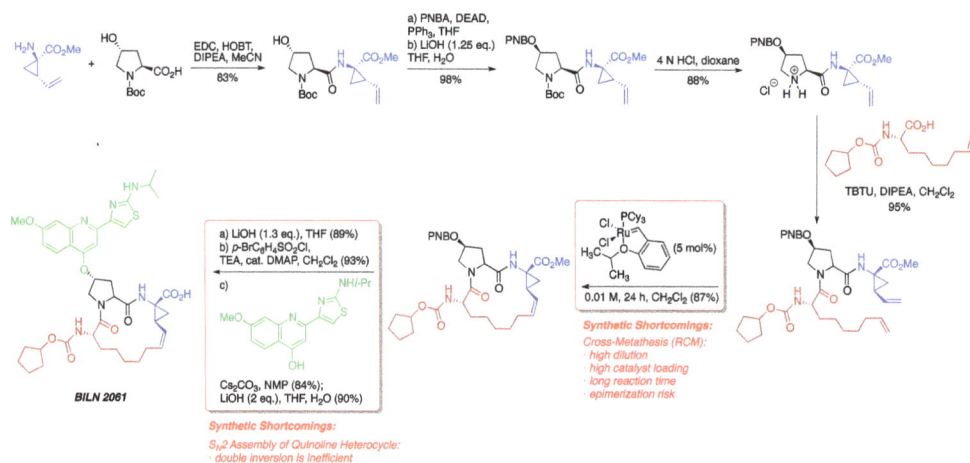

Analysis of the cross-methathesis reaction revealed that the conformation of the acyclic precursor had a profound impact on the formation of dimers and oligomers in the reaction mixture. By installing a Boc protecting group at the C-4 amide nitrogen, the Boehringer-Ingelheim chemists were able to shift the site of initiation from the vinylcyclopropane moiety to the nonenoic acid moiety, improving the rate of the intramolecular reaction and decreasing the risk of epimerization. Additionally, the catalyst employed was switched from the expensive 1st generation Hoveyda catalyst to the more reactive, less expensive Grela catalyst. These modifications allowed the process chemists to run the reaction at a standard reaction dilution of 0.1-0.2 M, given that the rates of competing dimerization and oligomerization reactions was so dramatically reduced.

Additionally, the process chemistry team envisioned a S_NAr strategy to install the quinoline heterocycle, instead of the S_N2 strategy that they had employed for the synthesis of BILN 2061. This modification prevented the need for inefficient double inversion by proceeding through retention of stereochemistry at the C-4 position of the hydroxyproline moiety.

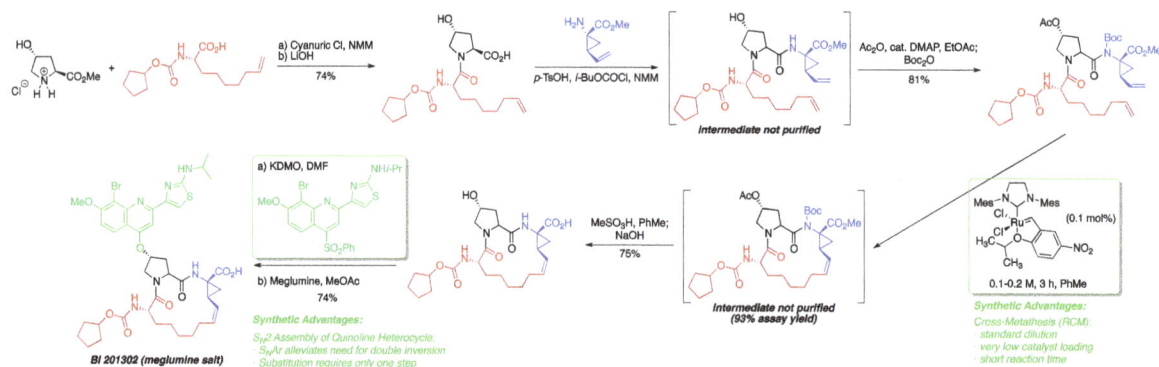

It is interesting to examine this case study from a VTO perspective. For the unoptimized cross-metathesis reaction using the Grela catalyst at 0.01 M diene, the reaction yield was determined to be 82 percent after a reaction and workup time of 48 hours. A 6-cubic meter reactor filled to 80% capacity afforded 35 kg of desired product. For the unoptimized reaction:

$$VTO = \frac{6 \text{ m}^3 \times 48 \text{ h}}{35 \text{ kg}} = 8.2 \text{ m}^3 \cdot \text{h / kg}$$

This VTO value was considered prohibitively high and a steep investment in a dedicated plant would have been necessary even before launching Phase III trials with this API, given its large projected annual demand. But after reaction development and optimization, the process team was able to improve the reaction yield to 93 percent after just 1 hour (plus 12 hours for workup and reactor cleaning time) at a diene concentration of 0.2 M. With these modifications, a 6-cubic meter reactor filled to 80% capacity afforded 799 kg of desired product. For this optimized reaction:

$$VTO = \frac{6 \text{ m}^3 \times 13 \text{ h}}{799 \text{ kg}} = 0.1 \text{ m}^3 \cdot \text{h} / \text{kg}$$

Thus, after optimization, this synthetic step became less costly in terms of equipment and time and more practical to perform in a standard manufacturing facility, eliminating the need for a costly investment in a new dedicated plant.

Transition-metal Catalysis and Organocatalysis

Biocatalysis and Enzymatic Engineering

Recently, large pharmaceutical process chemists have relied heavily on the development of enzymatic reactions to produce important chiral building blocks for API synthesis. Many varied classes of naturally occurring enzymes have been co-opted and engineered for process pharmaceutical chemistry applications. The widest range of applications come from ketoreductases and transaminases, but there are isolated examples from hydrolases, aldolases, oxidative enzymes, esterases and dehalogenases, among others.

One of the most prominent uses of biocatalysis in process chemistry today is in the synthesis of Januvia®, a DPP-4 inhibitor developed by Merck for the management of type II diabetes. The traditional process synthetic route involved a late-stage enamine formation followed by rhodium-catalyzed asymmetric hydrogenation to afford the API sitagliptin. This process suffered from a number of limitations, including the need to run the reaction under a high-pressure hydrogen environment, the high cost of a transition-metal catalyst, the difficult process of carbon treatment to remove trace amounts of catalyst and insufficient stereoselectivity, requiring a subsequent recrystallization step before final salt formation.

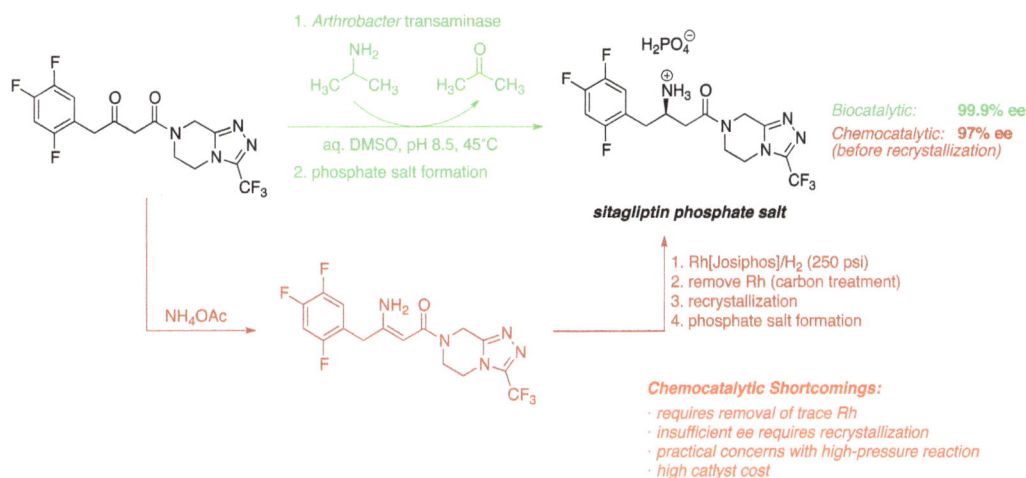

Merck's process chemistry department contracted Codexis, a medium-sized biocatalysis firm, to develop a large-scale biocatalytic reductive amination for the final step of its sitagliptin synthesis. Codexis engineered a transaminase enzyme from the bacteria Arthrobacter through 11 rounds of directed evolution. The engineered transaminase contained 27 individual point mutations and displayed activity four orders of magnitude greater than the parent enzyme. Additionally, the enzyme was engineered to handle high substrate concentrations (100 g/L) and to tolerate the organic solvents, reagents and byproducts of the transamination reaction. This biocatalytic route successfully avoided the limitations of the chemocatalyzed hydrogenation route: the requirements to run the reaction under high pressure, to remove excess catalyst by carbon treatment and to recrystallize the product due to insufficient enantioselectivity were obviated by the use of a biocatalyst. Merck and Codexis were awarded the Presidential Green Chemistry Challenge Award in 2010 for the development of this biocatalytic route toward Januvia®.

Continuous/Flow Manufacturing

In recent years, much progress has been made in the development and optimization of flow reactors for small-scale chemical synthesis (the Jamison Group at MIT and Ley Group at Cambridge University, among others, have pioneered efforts in this field). The pharmaceutical industry, however, has been slow to adopt this technology for large-scale synthetic operations. For certain reactions, however, continuous processing may possess distinct advantages over batch processing in terms of safety, quality and throughput.

A case study of particular interest involves the development of a fully continuous process by the process chemistry group at Eli Lilly and Company for an asymmetric hydrogenation to access a key intermediate in the synthesis of LY500307, a potent ERβ agonist that is entering clinical trials for the treatment of patients with schizophrenia, in addition to a regimen of standard antipsychotic medications. In this key synthetic step, a chiral rhodium-catalyst is used for the enantioselective reduction of a tetrasubstituted olefin. After extensive optimization, it was found that in order to reduce the catalyst loading to a commercially practical level, the reaction required hydrogen pressure up to 70 atm. The pressure limit of a standard chemical reactor is about 10 atm, although high-pressure batch reactors may be acquired at significant capital cost for reactions up to 100 atm. Especially for an API in the early stages of chemical development, such an investment clearly bears a large risk.

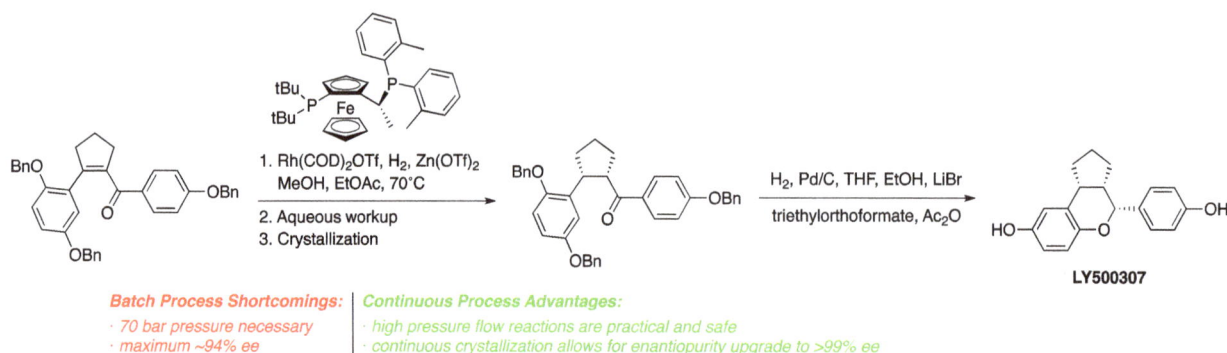

Batch Process Shortcomings:
· 70 bar pressure necessary
· maximum ~94% ee

Continuous Process Advantages:
· high pressure flow reactions are practical and safe
· continuous crystallization allows for enantiopurity upgrade to >99% ee

An additional concern was that the hydrogenation product has an unfavorable eutectic point, so it was impossible to isolate the crude intermediate in more than 94 percent ee by batch process. Because of this limitation, the process chemistry route toward LY500307 necessarily involved a kinetically controlled crystallization step after the hydrogenation to upgrade the enantiopurity of this penultimate intermediate to >99 percent ee.

The process chemistry team at Eli Lilly successfully developed a fully continuous process to this penultimate intermediate, including reaction, workup and kinetically controlled crystallization modules. An advantage of flow reactors is that high-pressure tubing can be utilized for hydrogenation and other hyperbaric reactions. Because the head space of a batch reactor is eliminated, however, many of the safety concerns associated with running high-pressure reactions are obviated by the use of a continuous process reactor. Additionally, a two-stage mixed suspension-mixed product removal (MSMPR) module was designed for the scalable, continuous, kinetically controlled crystallization of the product, so it was possible to isolate in >99 percent ee, eliminating the need for an additional batch crystallization step.

This continuous process afforded 144 kg of the key intermediate in 86 percent yield, comparable with a 90 percent isolated yield using the batch process. This 73-liter pilot-scale flow reactor (occupying less than $0.5 \ m^3$ space) achieved the same weekly throughput as theoretical batch processing in a 400-liter reactor. Therefore, the continuous flow process demonstrates advantages in safety, efficiency (eliminates the need for batch crystallization) and throughput, compared with a theoretical batch process.

Biochemistry

Biochemistry, sometimes called biological chemistry, is the study of chemical processes within and relating to living organisms. By controlling information flow through biochemical signaling and the flow of chemical energy through metabolism, biochemical processes give rise to the complexity of life. Over the last decades of the 20th century, biochemistry has become so successful at explaining living processes that now almost all areas of the life sciences from botany to medicine to genetics are engaged in biochemical research. Today, the main focus of pure biochemistry is on understanding how biological molecules give rise to the processes that occur within living cells, which in turn relates greatly to the study and understanding of tissues, organs, and whole organisms—that is, all of biology.

Biochemistry is closely related to molecular biology, the study of the molecular mechanisms by which genetic information encoded in DNA is able to result in the processes of life. Depending on the exact definition of the terms used, molecular biology can be thought of as a branch of biochemistry, or biochemistry as a tool with which to investigate and study molecular biology.

Much of biochemistry deals with the structures, functions and interactions of biological macromolecules, such as proteins, nucleic acids, carbohydrates and lipids, which provide the structure of cells and perform many of the functions associated with life. The chemistry of the cell also depends on the reactions of smaller molecules and ions. These can be inorganic, for example water and metal ions, or organic, for example the amino acids, which are used to synthesize proteins.

The mechanisms by which cells harness energy from their environment via chemical reactions are known as metabolism. The findings of biochemistry are applied primarily in medicine, nutrition, and agriculture. In medicine, biochemists investigate the causes and cures of diseases. In nutrition, they study how to maintain health and study the effects of nutritional deficiencies. In agriculture, biochemists investigate soil and fertilizers, and try to discover ways to improve crop cultivation, crop storage and pest control.

History

Gerty Cori and Carl Cori jointly won the Nobel Prize in 1947 for their discovery of the Cori cycle at RPMI.

At its broadest definition, biochemistry can be seen as a study of the components and composition of living things and how they come together to become life, and the history of biochemistry may therefore go back as far as the ancient Greeks. However, biochemistry as a specific scientific discipline has its beginning sometime in the 19th century, or a little earlier, depending on which aspect of biochemistry is being focused on. Some argued that the beginning of biochemistry may have been the discovery of the first enzyme, diastase (today called amylase), in 1833 by Anselme Payen, while others considered Eduard Buchner's first demonstration of a complex biochemical process alcoholic fermentation in cell-free extracts in 1897 to be the birth of biochemistry. Some might also point as its beginning to the influential 1842 work by Justus von Liebig, *Animal chemistry, or, Organic chemistry in its applications to physiology and pathology*, which presented a chemical theory of metabolism, or even earlier to the 18th century studies on fermentation and respiration by Antoine Lavoisier. Many other pioneers in the field who helped to uncover the layers of complexity of biochemistry have been proclaimed founders of modern biochemistry, for example Emil Fischer for his work on the chemistry of proteins, and F. Gowland Hopkins on enzymes and the dynamic nature of biochemistry.

The term "biochemistry" itself is derived from a combination of biology and chemistry. In 1877, Felix Hoppe-Seyler used the term (*biochemie* in German) as a synonym for physiological chemistry in the foreword to the first issue of *Zeitschrift für Physiologische Chemie* where he argued for the setting up of institutes dedicated to this field of study. The German chemist Carl Neuberg however is often cited to have coined the word in 1903, while some credited it to Franz Hofmeister.

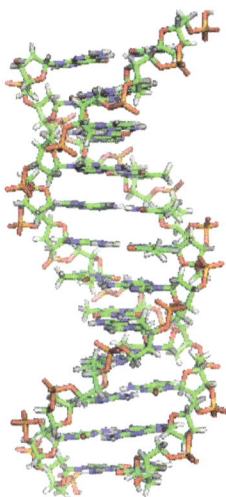

DNA structure (1D65)

It was once generally believed that life and its materials had some essential property or substance (often referred to as the "vital principle") distinct from any found in non-living matter, and it was thought that only living beings could produce the molecules of life. Then, in 1828, Friedrich Wöhler published a paper on the synthesis of urea, proving that organic compounds can be created artificially. Since then, biochemistry has advanced, especially since the mid-20th century, with the development of new techniques such as chromatography, X-ray diffraction, dual polarisation interferometry, NMR spectroscopy, radioisotopic labeling, electron microscopy, and molecular dynamics simulations. These techniques allowed for the discovery and detailed analysis of many molecules and metabolic pathways of the cell, such as glycolysis and the Krebs cycle (citric acid cycle).

Another significant historic event in biochemistry is the discovery of the gene and its role in the transfer of information in the cell. This part of biochemistry is often called molecular biology. In the 1950s, James D. Watson, Francis Crick, Rosalind Franklin, and Maurice Wilkins were instrumental in solving DNA structure and suggesting its relationship with genetic transfer of information. In 1958, George Beadle and Edward Tatum received the Nobel Prize for work in fungi showing that one gene produces one enzyme. In 1988, Colin Pitchfork was the first person convicted of murder with DNA evidence, which led to the growth of forensic science. More recently, Andrew Z. Fire and Craig C. Mello received the 2006 Nobel Prize for discovering the role of RNA interference (RNAi), in the silencing of gene expression.

Starting Materials: The Chemical Elements of Life

Around two dozen of the 92 naturally occurring chemical elements are essential to various kinds of biological life. Most rare elements on Earth are not needed by life (exceptions being selenium and iodine), while a few common ones (aluminum and titanium) are not used. Most organisms share element needs, but there are a few differences between plants and animals. For example, ocean algae use bromine, but land plants and animals seem to need none. All animals require sodium, but some plants do not. Plants need boron and silicon, but animals may not (or may need ultra-small amounts).

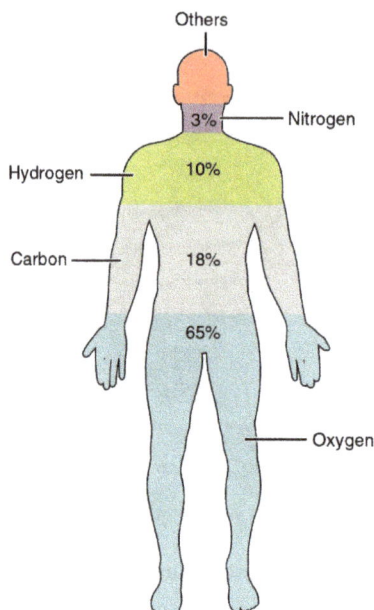

The main elements that compose the human body are shown from most abundant (by mass) to least abundant.

Just six elements—carbon, hydrogen, nitrogen, oxygen, calcium, and phosphorus—make up almost 99% of the mass of living cells, including those in the human body. In addition to the six major elements that compose most of the human body, humans require smaller amounts of possibly 18 more.

Biomolecules

The four main classes of molecules in biochemistry (often called biomolecules) are carbohydrates, lipids, proteins, and nucleic acids. Many biological molecules are polymers: in this terminology, monomers are relatively small micromolecules that are linked together to create large macromolecules known as polymers. When monomers are linked together to synthesize a biological polymer, they undergo a process called dehydration synthesis. Different macromolecules can assemble in larger complexes, often needed for biological activity.

Carbohydrates

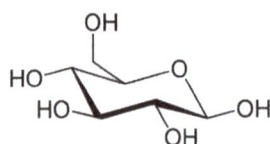

Glucose, a monosaccharide

The function of carbohydrates includes energy storage and providing structure. Sugars are carbohydrates, but not all carbohydrates are sugars. There are more carbohydrates on Earth than any other known type of biomolecule; they are used to store energy and genetic information, as well as play important roles in cell to cell interactions and communications.

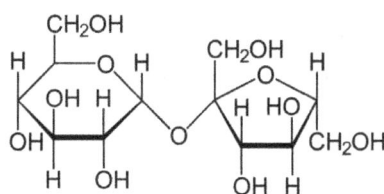

A molecule of sucrose (glucose + fructose), a disaccharide

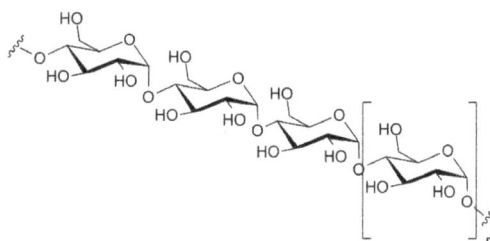

Amylose, a polysaccharide made up of several thousand glucose units

The simplest type of carbohydrate is a monosaccharide, which among other properties contains carbon, hydrogen, and oxygen, mostly in a ratio of 1:2:1 (generalized formula $C_nH_{2n}O_n$, where n is at least 3). Glucose ($C_6H_{12}O_6$) is one of the most important carbohydrates, others include fructose ($C_6H_{12}O_6$), the sugar commonly associated with the sweet taste of fruits,[a] and deoxyribose ($C_5H_{10}O_4$).

A monosaccharide can switch from the acyclic (open-chain) form to a cyclic form, through a nucleophilic addition reaction between the carbonyl group and one of the hydroxyls of the same molecule. The reaction creates a ring of carbon atoms closed by one bridging oxygen atom. The resulting molecule has an hemiacetal or hemiketal group, depending on whether the linear form was an aldose or a ketose. The reaction is easily reversed, yielding the original open-chain form.

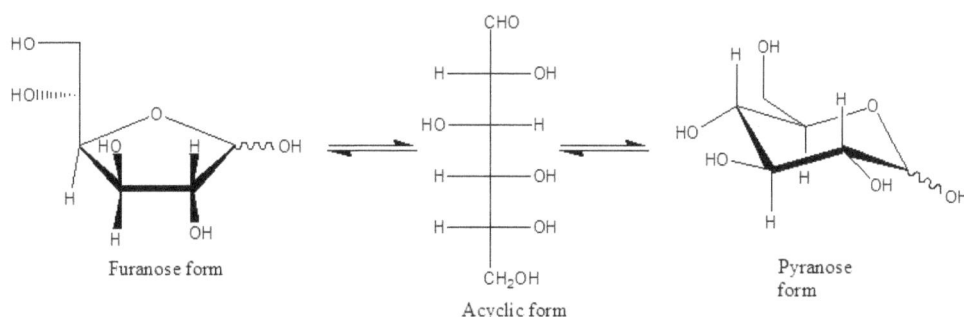

Conversion between the furanose, acyclic, and pyranose forms of D-glucose.

In these cyclic forms, the ring usually has 5 or 6 atoms. These forms are called furanoses and pyranoses, respectively — by analogy with furan and pyran, the simplest compounds with the same carbon-oxygen ring (although they lack the double bonds of these two molecules). For example, the aldohexose glucose may form a hemiacetal linkage between the hydroxyl on carbon 1 and the oxygen on carbon 4, yielding a molecule with a 5-membered ring, called glucofuranose. The same reaction can take place between carbons 1 and 5 to form a molecule with a 6-membered ring, called glucopyranose. Cyclic forms with a 7-atom ring (the same of oxepane), rarely encountered, are called heptoses.

When two monosaccharides undergo dehydration synthesis whereby a molecule of water is released, as two hydrogen atoms and one oxygen atom are lost from the two monosaccharides. The new molecule, consisting of two monosaccharides, is called a *disaccharide* and is conjoined together by a glycosidic or ether bond. The reverse reaction can also occur, using a molecule of water to split up a disaccharide and break the glycosidic bond; this is termed *hydrolysis*. The most well-known disaccharide is sucrose, ordinary sugar (in scientific contexts, called *table sugar* or *cane sugar* to differentiate it from other sugars). Sucrose consists of a glucose molecule and a fructose molecule joined together. Another important disaccharide is lactose, consisting of a glucose molecule and a galactose molecule. As most humans age, the production of lactase, the enzyme that hydrolyzes lactose back into glucose and galactose, typically decreases. This results in lactase deficiency, also called *lactose intolerance*.

When a few (around three to six) monosaccharides are joined, it is called an *oligosaccharide* (*oligo-* meaning "few"). These molecules tend to be used as markers and signals, as well as having some other uses. Many monosaccharides joined together make a polysaccharide. They can be joined together in one long linear chain, or they may be branched. Two of the most common polysaccharides are cellulose and glycogen, both consisting of repeating glucose monomers. Examples are *Cellulose* which is an important structural component of plant's cell walls, and *glycogen*, used as a form of energy storage in animals.

Sugar can be characterized by having reducing or non-reducing ends. A reducing end of a carbohydrate is a carbon atom that can be in equilibrium with the open-chain aldehyde (aldose) or keto form (ketose). If the joining of monomers takes place at such a carbon atom, the free hydroxy group of the pyranose or furanose form is exchanged with an OH-side-chain of another sugar, yielding a full acetal. This prevents opening of the chain to the aldehyde or keto form and renders the modified residue non-reducing. Lactose contains a reducing end at its glucose moiety, whereas the galactose moiety form a full acetal with the C4-OH group of glucose. Saccharose does not have a reducing end because of full acetal formation between the aldehyde carbon of glucose (C1) and the keto carbon of fructose (C2).

Lipids

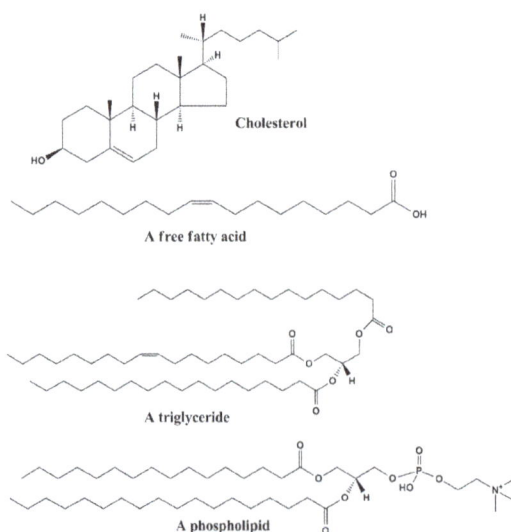

Structures of some common lipids. At the top are cholesterol and oleic acid. The middle structure is a triglyceride composed of oleoyl, stearoyl, and palmitoyl chains attached to a glycerol backbone. At the bottom is the common phospholipid, phosphatidylcholine.

Lipids comprises a diverse range of molecules and to some extent is a catchall for relatively water-insoluble or nonpolar compounds of biological origin, including waxes, fatty acids, fatty-acid derived phospholipids, sphingolipids, glycolipids, and terpenoids (e.g., retinoids and steroids). Some lipids are linear aliphatic molecules, while others have ring structures. Some are aromatic, while others are not. Some are flexible, while others are rigid.

Lipids are usually made from one molecule of glycerol combined with other molecules. In triglycerides, the main group of bulk lipids, there is one molecule of glycerol and three fatty acids. Fatty acids are considered the monomer in that case, and may be saturated (no double bonds in the carbon chain) or unsaturated (one or more double bonds in the carbon chain).

Most lipids have some polar character in addition to being largely nonpolar. In general, the bulk of their structure is nonpolar or hydrophobic ("water-fearing"), meaning that it does not interact well with polar solvents like water. Another part of their structure is polar or hydrophilic ("water-loving") and will tend to associate with polar solvents like water. This makes them amphiphilic molecules (having both hydrophobic and hydrophilic portions). In the case of cholesterol, the polar group is a mere -OH (hydroxyl or alcohol). In the case of phospholipids, the polar groups are considerably larger and more polar, as described below.

Lipids are an integral part of our daily diet. Most oils and milk products that we use for cooking and eating like butter, cheese, ghee etc., are composed of fats. Vegetable oils are rich in various polyunsaturated fatty acids (PUFA). Lipid-containing foods undergo digestion within the body and are broken into fatty acids and glycerol, which are the final degradation products of fats and lipids. Lipids, especially phospholipids, are also used in various pharmaceutical products, either as co-solubilisers (e.g., in parenteral infusions) or else as drug carrier components (e.g., in a liposome or transfersome).

Proteins

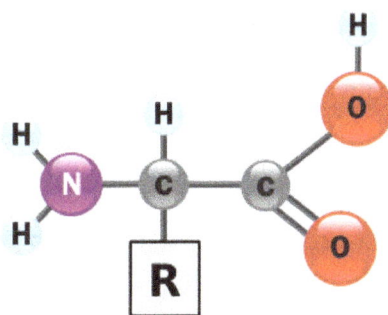

The general structure of an α-amino acid, with the amino group on the left and the carboxyl group on the right.

Proteins are very large molecules – macro-biopolymers – made from monomers called amino acids. An amino acid consists of a carbon atom bound to four groups. One is an amino group, $-NH_2$, and one is a carboxylic acid group, $-COOH$ (although these exist as $-NH_3^+$ and $-COO^-$ under physiologic conditions). The third is a simple hydrogen atom. The fourth is commonly denoted "$-R$" and is different for each amino acid. There are 20 standard amino acids, each containing a carboxyl group, an amino group, and a side-chain (known as an "R" group). The "R" group is what makes each amino acid different, and the properties of the side-chains greatly influence the overall three-dimensional conformation of a protein. Some amino acids have functions by themselves or in a modified form; for instance, glutamate functions as an important neurotransmitter. Amino acids can be joined via a peptide bond. In this dehydration synthesis, a water molecule is removed and the peptide bond connects the nitrogen of one amino acid's amino group to the carbon of the other's carboxylic acid group. The resulting molecule is called a *dipeptide*, and short stretches of amino acids (usually, fewer than thirty) are called *peptides* or polypeptides. Longer stretches merit the title *proteins*. As an example, the important blood serum protein albumin contains 585 amino acid residues.

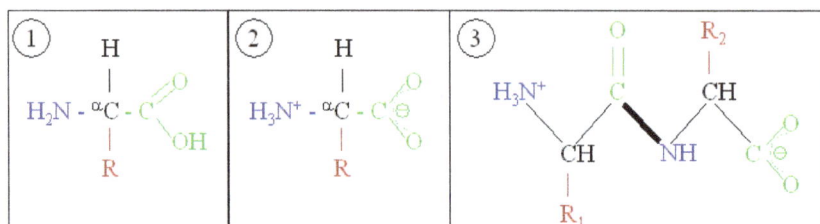

Generic amino acids (1) in neutral form, (2) as they exist physiologically, and (3) joined together as a dipeptide.

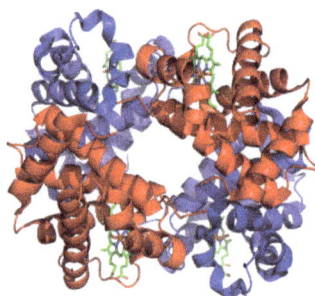

A schematic of hemoglobin. The red and blue ribbons represent the protein globin; the green structures are the heme groups.

Some proteins perform largely structural roles. For instance, movements of the proteins actin and myosin ultimately are responsible for the contraction of skeletal muscle. One property many proteins have is that they specifically bind to a certain molecule or class of molecules—they may be *extremely* selective in what they bind. Antibodies are an example of proteins that attach to one specific type of molecule. In fact, the enzyme-linked immunosorbent assay (ELISA), which uses antibodies, is one of the most sensitive tests modern medicine uses to detect various biomolecules. Probably the most important proteins, however, are the enzymes. Virtually every reaction in a living cell requires an enzyme to lower the activation energy of the reaction. These molecules recognize specific reactant molecules called *substrates*; they then catalyze the reaction between them. By lowering the activation energy, the enzyme speeds up that reaction by a rate of 10^{11} or more; a reaction that would normally take over 3,000 years to complete spontaneously might take less than a second with an enzyme. The enzyme itself is not used up in the process, and is free to catalyze the same reaction with a new set of substrates. Using various modifiers, the activity of the enzyme can be regulated, enabling control of the biochemistry of the cell as a whole.

The structure of proteins is traditionally described in a hierarchy of four levels. The primary structure of a protein simply consists of its linear sequence of amino acids; for instance, "alanine-glycine-tryptophan-serine-glutamate-asparagine-glycine-lysine-...". Secondary structure is concerned with local morphology (morphology being the study of structure). Some combinations of amino acids will tend to curl up in a coil called an α-helix or into a sheet called a β-sheet; some α-helixes can be seen in the hemoglobin schematic above. Tertiary structure is the entire three-dimensional shape of the protein. This shape is determined by the sequence of amino acids. In fact, a single change can change the entire structure. The alpha chain of hemoglobin contains 146 amino acid residues; substitution of the glutamate residue at position 6 with a valine residue changes the behavior of hemoglobin so much that it results in sickle-cell disease. Finally, quaternary structure is concerned with the structure of a protein with multiple peptide subunits, like hemoglobin with its four subunits. Not all proteins have more than one subunit.

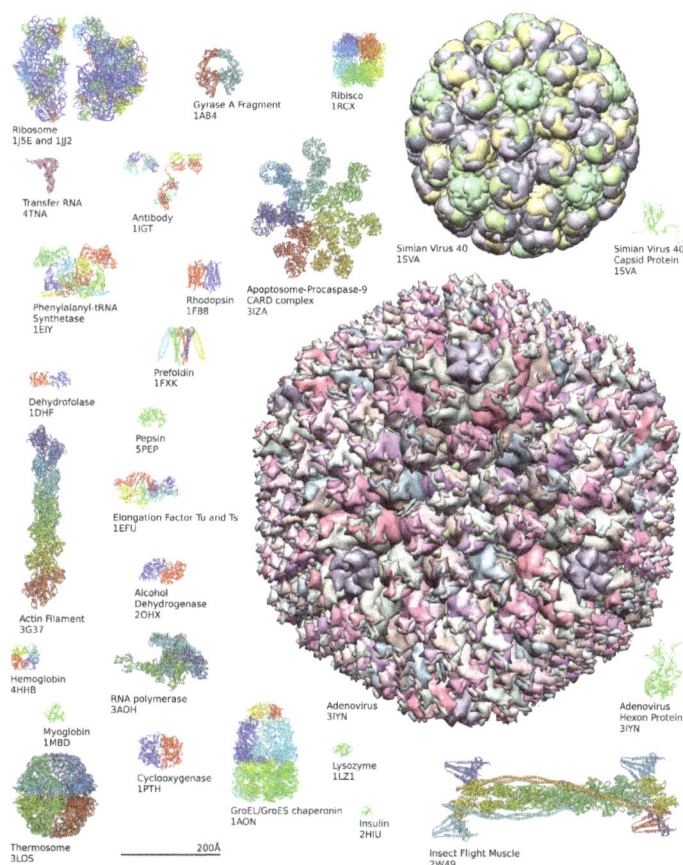

Examples of protein structures from the Protein Data Bank

Ingested proteins are usually broken up into single amino acids or dipeptides in the small intestine, and then absorbed. They can then be joined to make new proteins. Intermediate products of glycolysis, the citric acid cycle, and the pentose phosphate pathway can be used to make all twenty amino acids, and most bacteria and plants possess all the necessary enzymes to synthesize them. Humans and other mammals, however, can synthesize only half of them. They cannot synthesize isoleucine, leucine, lysine, methionine, phenylalanine, threonine, tryptophan, and valine. These are the essential amino acids, since it is essential to ingest them. Mammals do possess the enzymes to synthesize alanine, asparagine, aspartate, cysteine, glutamate, glutamine, glycine, proline, serine, and tyrosine, the nonessential amino acids. While they can synthesize arginine and histidine, they cannot produce it in sufficient amounts for young, growing animals, and so these are often considered essential amino acids.

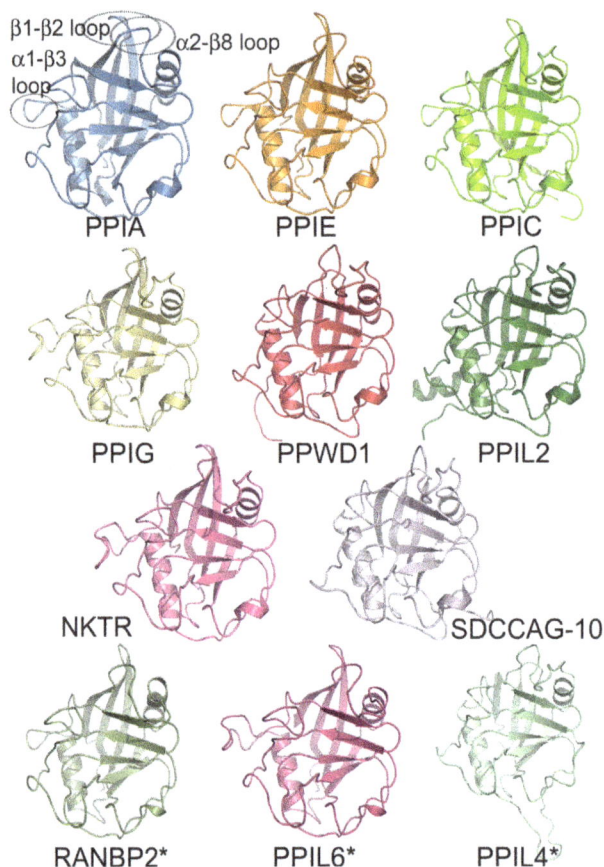

Members of a protein family, as represented by the structures of the isomerase domains.

If the amino group is removed from an amino acid, it leaves behind a carbon skeleton called an α-keto acid. Enzymes called transaminases can easily transfer the amino group from one amino acid (making it an α-keto acid) to another α-keto acid (making it an amino acid). This is important in the biosynthesis of amino acids, as for many of the pathways, intermediates from other biochemical pathways are converted to the α-keto acid skeleton, and then an amino group is added, often via transamination. The amino acids may then be linked together to make a protein.

A similar process is used to break down proteins. It is first hydrolyzed into its component amino acids. Free ammonia (NH_3), existing as the ammonium ion (NH_4^+) in blood, is toxic to life forms. A suitable method for excreting it must therefore exist. Different tactics have evolved in different animals, depending on the animals' needs. Unicellular organisms, of course, simply release the ammonia into the environment. Likewise, bony fish can release the ammonia into the water where it is quickly diluted. In general, mammals convert the ammonia into urea, via the urea cycle.

In order to determine whether two proteins are related, or in other words to decide whether they are homologous or not, scientists use sequence-comparison methods. Methods like sequence alignments and structural alignments are powerful tools that help scientists identify homologies between related molecules. The relevance of finding homologies among proteins goes beyond forming an evolutionary pattern of protein families. By finding how similar two protein sequences are, we acquire knowledge about their structure and therefore their function.

Nucleic Acids

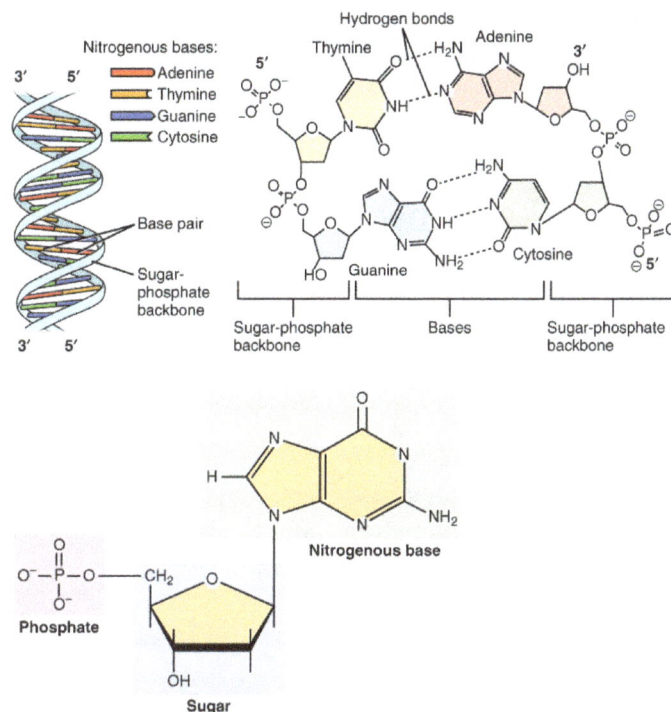

The structure of deoxyribonucleic acid (DNA), the picture shows the monomers being put together.

Nucleic acids, so called because of its prevalence in cellular nuclei, is the generic name of the family of biopolymers. They are complex, high-molecular-weight biochemical macromolecules that can convey genetic information in all living cells and viruses. The monomers are called nucleotides, and each consists of three components: a nitrogenous heterocyclic base (either a purine or a pyrimidine), a pentose sugar, and a phosphate group.

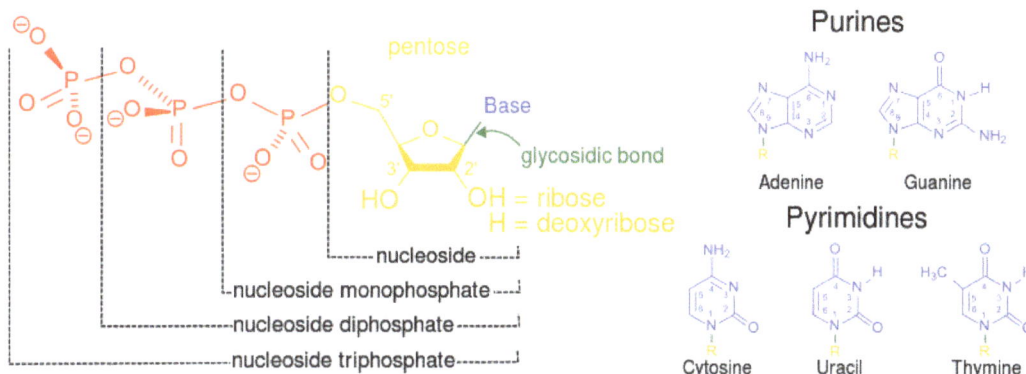

Structural elements of common nucleic acid constituents. Because they contain at least one phosphate group, the compounds marked *nucleoside monophosphate*, *nucleoside diphosphate* and *nucleoside triphosphate* are all nucleotides (not simply phosphate-lacking nucleosides).

The most common nucleic acids are deoxyribonucleic acid (DNA) and ribonucleic acid (RNA). The phosphate group and the sugar of each nucleotide bond with each other to form the backbone of the nucleic acid, while the sequence of nitrogenous bases stores the information. The most common nitrogenous bases are adenine, cytosine, guanine, thymine, and uracil. The nitrogenous bases of each strand of a nucleic acid will form hydrogen bonds with certain other nitrogenous bases in a complementary strand of nucleic acid (similar to a zipper). Adenine binds with thymine and uracil; Thymine binds only with adenine; and cytosine and guanine can bind only with one another.

Aside from the genetic material of the cell, nucleic acids often play a role as second messengers, as well as forming the base molecule for adenosine triphosphate (ATP), the primary energy-carrier molecule found in all living organisms. Also, the nitrogenous bases possible in the two nucleic acids are different: adenine, cytosine, and guanine occur in both RNA and DNA, while thymine occurs only in DNA and uracil occurs in RNA.

Metabolism

Carbohydrates as Energy Source

Glucose is the major energy source in most life forms. For instance, polysaccharides are broken down into their monomers (glycogen phosphorylase removes glucose residues from glycogen). Disaccharides like lactose or sucrose are cleaved into their two component monosaccharides.

Glycolysis (Anaerobic)

Glucose is mainly metabolized by a very important ten-step pathway called glycolysis, the net result of which is to break down one molecule of glucose into two molecules of pyruvate. This also produces a net two molecules of ATP, the energy currency of cells, along with two reducing equivalents of converting NAD+ (nicotinamide adenine dinucleotide:oxidised form) to NADH (nicotinamide adenine dinucleotide:reduced form). This does not require oxygen; if no oxygen is available (or the cell cannot use oxygen), the NAD is restored by converting the pyruvate to lactate (lactic acid) (e.g., in humans) or to ethanol plus carbon dioxide (e.g., in yeast). Other monosaccharides like galactose and fructose can be converted into intermediates of the glycolytic pathway.

Aerobic

In aerobic cells with sufficient oxygen, as in most human cells, the pyruvate is further metabolized. It is irreversibly converted to acetyl-CoA, giving off one carbon atom as the waste product carbon dioxide, generating another reducing equivalent as NADH. The two molecules acetyl-CoA (from one molecule of glucose) then enter the citric acid cycle, producing two more molecules of ATP, six more NADH molecules and two reduced (ubi)quinones (via $FADH_2$ as enzyme-bound cofactor), and releasing the remaining carbon atoms as carbon dioxide. The produced NADH and quinol molecules then feed into the enzyme complexes of the respiratory chain, an electron transport system transferring the electrons ultimately to oxygen and conserving the released energy in the form of a proton gradient over a membrane (inner mitochondrial membrane in eukaryotes). Thus, oxygen is reduced to water and the original electron acceptors NAD+ and quinone are regenerated. This is why humans breathe in oxygen and breathe out carbon dioxide. The energy released from transferring the electrons from high-energy states in NADH and quinol is conserved first as proton gradient and converted to ATP via ATP synthase. This generates an additional 28 molecules of ATP (24 from the 8 NADH + 4 from the 2 quinols), totaling to 32 molecules of ATP conserved per degraded glucose (two from glycolysis + two from the citrate cycle). It is clear that using oxygen to completely oxidize glucose provides an organism with far more energy than any oxygen-independent metabolic feature, and this is thought to be the reason why complex life appeared only after Earth's atmosphere accumulated large amounts of oxygen.

Gluconeogenesis

In vertebrates, vigorously contracting skeletal muscles (during weightlifting or sprinting, for example) do not receive enough oxygen to meet the energy demand, and so they shift to anaerobic metabolism, converting glucose to lactate. The liver regenerates the glucose, using a process called gluconeogenesis. This process is not quite the opposite of glycolysis, and actually requires three times the amount of energy gained from glycolysis (six molecules of ATP are used, compared to the two gained in glycolysis). Analogous to the above reactions, the glucose produced can then undergo glycolysis in tissues that need energy, be stored as glycogen (or starch in plants), or be converted to other monosaccharides or joined into di- or oligosaccharides. The combined pathways of glycolysis during exercise, lactate's crossing via the bloodstream to the liver, subsequent gluconeogenesis and release of glucose into the bloodstream is called the Cori cycle.

Relationship to other "Molecular-Scale" Biological Sciences

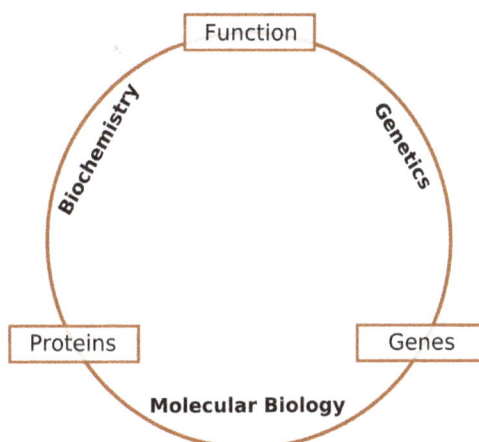

Schematic relationship between biochemistry, genetics, and molecular biology.

Researchers in biochemistry use specific techniques native to biochemistry, but increasingly combine these with techniques and ideas developed in the fields of genetics, molecular biology and biophysics. There has never been a hard-line among these disciplines in terms of content and technique. Today, the terms *molecular biology* and *biochemistry* are nearly interchangeable. The following figure is a schematic that depicts one possible view of the relationship between the fields:

- *Biochemistry* is the study of the chemical substances and vital processes occurring in living organisms. Biochemists focus heavily on the role, function, and structure of biomolecules. The study of the chemistry behind biological processes and the synthesis of biologically active molecules are examples of biochemistry.

- *Genetics* is the study of the effect of genetic differences on organisms. Often this can be inferred by the absence of a normal component (e.g., one gene). The study of "mutants" –

organisms with a changed gene that leads to the organism being different with respect to the so-called "wild type" or normal phenotype. Genetic interactions (epistasis) can often confound simple interpretations of such "knock-out" or "knock-in" studies.

- *Molecular biology* is the study of molecular underpinnings of the process of replication, transcription and translation of the genetic material. The central dogma of molecular biology where genetic material is transcribed into RNA and then translated into protein, despite being an oversimplified picture of molecular biology, still provides a good starting point for understanding the field. This picture, however, is undergoing revision in light of emerging novel roles for RNA.

- *Chemical biology* seeks to develop new tools based on small molecules that allow minimal perturbation of biological systems while providing detailed information about their function. Further, chemical biology employs biological systems to create non-natural hybrids between biomolecules and synthetic devices (for example emptied viral capsids that can deliver gene therapy or drug molecules).

References

- Vogel, A.I., Tatchell, A.R., Furnis, B.S., Hannaford, A.J. and P.W.G. Smith. Vogel's Textbook of Practical Organic Chemistry, 5th Edition. Prentice Hall, 1996. ISBN 0-582-46236-3.

- Desai, A. A. (2011). "Sitagliptin manufacture: a compelling tale of green chemistry, process intensification, and industrial asymmetric catalysis". Angew. Chem. Int. Ed. 50: 1974–1976. doi:10.1002/anie.201007051.

5

Various Chemical Reactions Involved in Green Chemistry

Condensation reaction is a chemical reaction and in this reaction, two molecules combine to form a large molecule. Other chemical reactions explained in the section are aldol condensation, Arndt-Eistert synthesis, Baeyer-Villiger oxidation, Dakin reaction, Simmons-Smith reaction, Heck reaction etc. The aspects elucidated in this section are of vital importance, and provides a better understanding of chemical reactions.

Condensation Reaction

A condensation reaction, is a chemical reaction in which two molecules or moieties, often functional groups, combine to form a larger molecule, together with the loss of a small molecule. Possible small molecules that are lost include water, hydrogen chloride, methanol, or acetic acid, but most commonly in a biological reaction it is water.

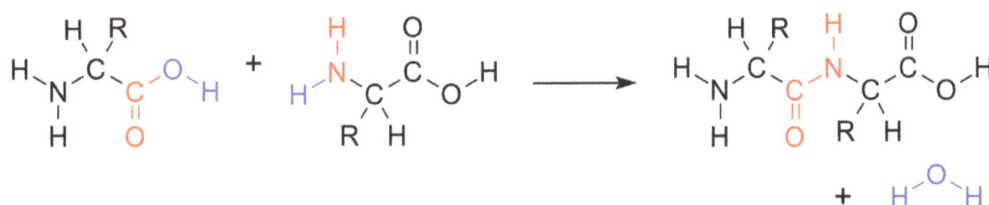

The condensation of two amino acids to form a peptide bond (red) with expulsion of water (blue)

When two separate molecules react, the condensation is termed intermolecular. A simple example is the condensation of two amino acids to form the peptide bond characteristic of proteins. This reaction example is the opposite of hydrolysis, which splits a chemical entity into two parts through the action of the polar water molecule, which itself splits into hydroxide and hydrogen ions. Hence energy is required to form chemical bonds via condensation.

If the union is between atoms or groups of the same molecule, the reaction is termed intramolecular condensation, and in many cases leads to ring formation. An example is the Dieckmann condensation, in which the two ester groups of a single diester molecule react with each other to lose a small alcohol molecule and form a β-ketoester product.

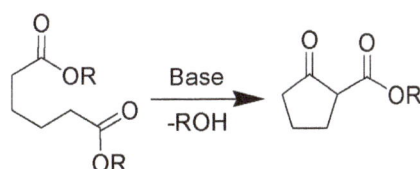

Dieckmann condensation reaction

Mechanism

Many condensation reactions follow a nucleophilic acyl substitution or an aldol condensation reaction mechanism. Other condensations, such as the acyloin condensation are triggered by radical or single electron transfer conditions.

Condensation Reactions in Polymer Chemistry

In one type of polymerization reaction, a series of condensation steps take place whereby monomers or monomer chains add to each other to form longer chains called polymers. This is termed 'condensation polymerization' or 'step-growth polymerization', and occurs for example in the synthesis of polyesters or nylons. It may be either a homopolymerization of a single monomer A-B with two different end groups that condense or a copolymerization of two co-monomers A-A and B-B. Small molecules are usually liberated in these condensation steps, in contrast to polyaddition reactions with no liberation of small molecules.

In general, condensation polymers form more slowly than addition polymers, often requiring heat. They are generally lower in molecular weight. Monomers are consumed early in the reaction; the terminal functional groups remain active throughout and short chains combine to form longer chains. A high conversion rate is required to achieve high molecular weights as per Carothers' equation.

Bifunctional monomers lead to linear chains, and therefore thermoplastic polymers, but, when the monomer functionality exceeds two, the product is a branched chain that may yield a thermoset polymer.

Applications

This type of reaction is used as a basis for the making of many important polymers, for example: nylon, polyester, and other condensation polymers and various epoxies. It is also the basis for the laboratory formation of silicates and polyphosphates. The reactions that form acid anhydrides from their constituent acids are typically condensation reactions.

Many biological transformations are condensation reactions. Polypeptide synthesis, polyketide synthesis, terpene syntheses, phosphorylation, and glycosylations are a few examples of this type of reaction. A large number of such reactions are used in synthetic organic chemistry. Other examples include:

- Acyloin condensation
- Aldol condensation
- Benzoin condensation (this is not technically a condensation, but is called so for historical reasons)
- Claisen condensation
- Claisen–Schmidt condensation
- Darzens condensation (glycidic ester condensation)

- Dieckmann condensation

- Guareschi–Thorpe condensation

- Knoevenagel condensation

- Michael condensation

- Pechmann condensation

- Rap–Stoermer condensation

- Self-condensation or symmetrical aldol condensation

- Ziegler condensation

Acyloin Condensation

Acyloin condensation is a reductive coupling of two carboxylic esters using metallic sodium to yield an α-hydroxyketone, also known as an acyloin.

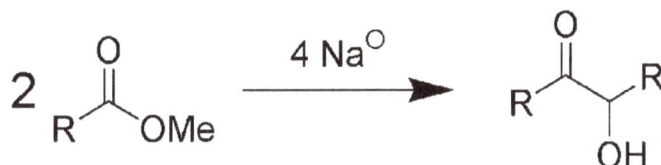

The reaction is most successful when R is aliphatic and inert. The reaction is performed in aprotic solvents with a high boiling point, such as benzene and toluene. The use of protic solvents results in the Bouveault-Blanc reduction of the separate esters rather than condensation. Depending on ring size and steric properties, but independent from high dilution, the acyloin condensation of diesters favours intramolecular cyclisation over intermolecular polymerisation.

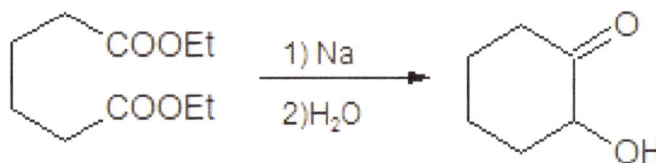

Mechanism

The mechanism consists of four steps:

(1) Oxidative ionization of two sodium atoms on the double bond of two ester molecules.

(2) Free radical coupling between two molecules of the homolytic ester derivative (A Wurtz type coupling). Alkoxy-eliminations in both sides occur, producing a 1,2-diketone.

(3) Oxidative ionization of two sodium atoms on both diketone double bonds. The sodium enodiolate is formed.

(4) Neutralization with water to form the enodiol, which tautomerizes to acyloin.

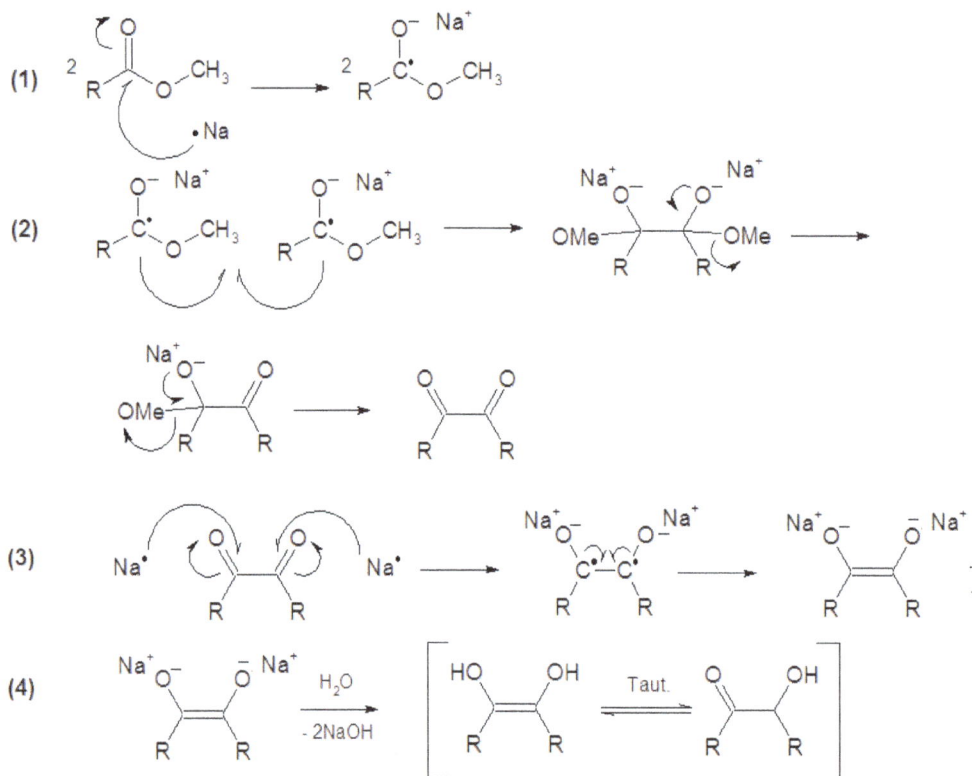

Variations

Rühlmann-method

The method according to Rühlmann employs trimethylchlorosilane as a trapping reagent; by this, competing reactions are efficiently subdued. Generally, yields increase considerably. The hydrolytic cleavage of the silylether gives the acyloin. To achieve a mild cleavage methanol can be used in several cases.

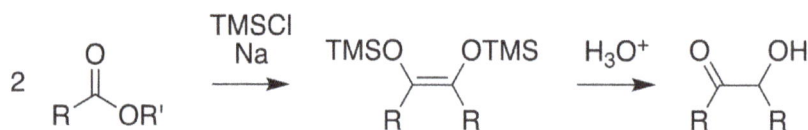

Usually toluene, dioxane, tetrahydrofuran or acyclic dialkylethers are employed as solvents. Advantageously also N-methyl-morpholine has been used. It allowed in some cases a successful reaction, in which otherwise the reaction failed in less polar media.

Aldol Condensation

An aldol condensation is a condensation reaction in organic chemistry in which an enol or an enolate ion reacts with a carbonyl compound to form a β-hydroxyaldehyde or β-hydroxyketone, followed by dehydration to give a conjugated enone.

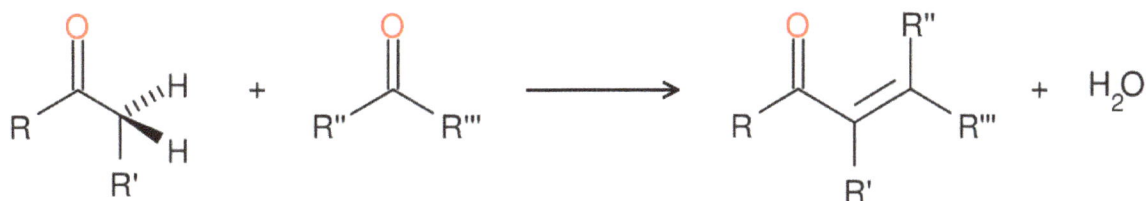

Aldol condensations are important in organic synthesis, because they provide a good way to form carbon–carbon bonds. For example, the Robinson annulation reaction sequence features an aldol condensation; the Wieland-Miescher ketone product is an important starting material for many organic syntheses. Aldol condensations are also commonly discussed in university level organic chemistry classes as a good bond-forming reaction that demonstrates important reaction mechanisms. In its usual form, it involves the nucleophilic addition of a ketone enolate to an aldehyde to form a β-hydroxy ketone, or "aldol" (aldehyde + alcohol), a structural unit found in many naturally occurring molecules and pharmaceuticals.

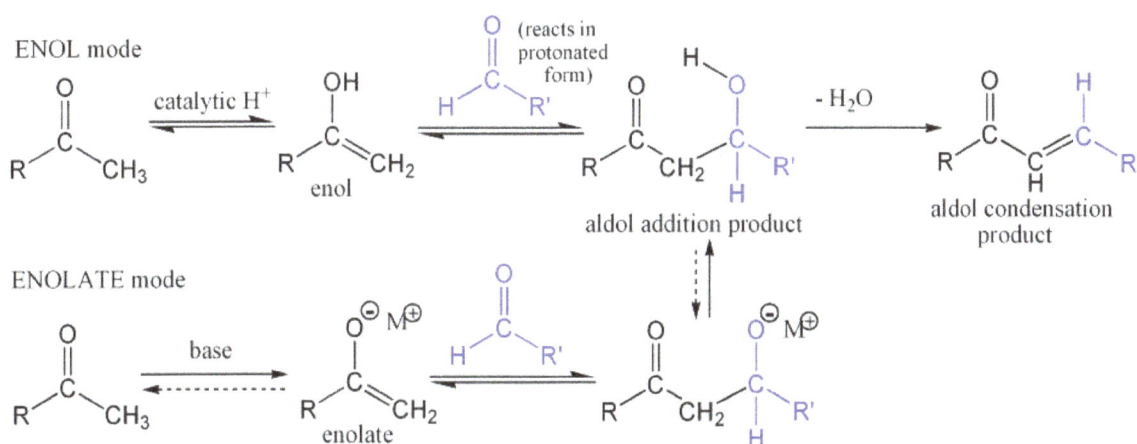

The name aldol condensation is also commonly used, especially in biochemistry, to refer to just the first (addition) stage of the process—the aldol reaction itself—as catalyzed by aldolases. However, the aldol reaction is not formally a condensation reaction because it does not involve the loss of a small molecule.

The reaction between an aldehyde/ketone and an aromatic carbonyl compound lacking an alpha-hydrogen (cross aldol condensation) is called the Claisen-Schmidt condensation. This reaction is named after two of its pioneering investigators Rainer Ludwig Claisen and J. G. Schmidt, who independently published on this topic in 1880 and 1881. An example is the synthesis of dibenzylideneacetone. Quantitative yields in Claisen-Schmidt reactions have been reported in the absence of solvent using sodium hydroxide as the base and plus benzaldehydes.

Mechanism

The first part of this reaction is an aldol reaction, the second part a dehydration—an elimination reaction (Involves removal of a water molecule or an alcohol molecule). Dehydration may be accompanied by decarboxylation when an activated carboxyl group is present. The aldol addition product can be dehydrated via two mechanisms; a strong base like potassium *t*-butoxide, potassium hydroxide or sodium hydride in an enolate mechanism, or in an acid-catalyzed enol mechanism. Depending on the nature of the desired product, the aldol condensation may be carried out under two broad types of conditions: kinetic control or thermodynamic control.

Base catalyzed aldol reaction (shown using ⁻OCH$_3$ as base)

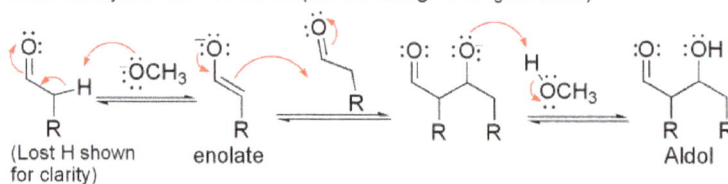

Base catalyzed dehydration (sometimes written as a single step)

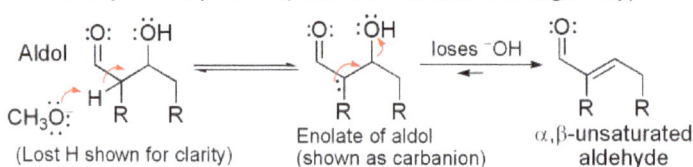

Acid catalyzed aldol reaction

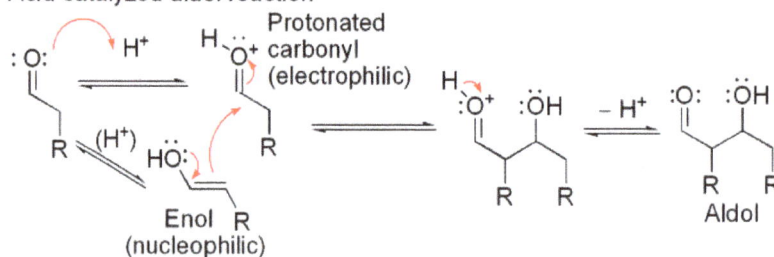

Acid catalyzed dehydration

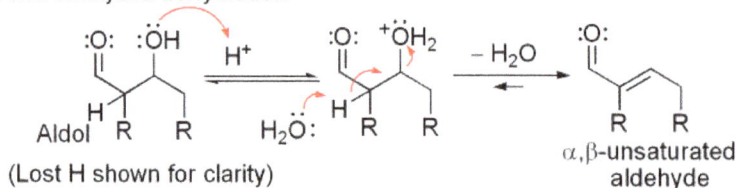

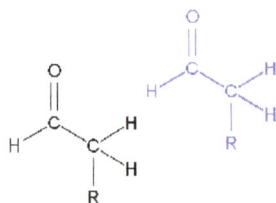

animation, base catalyzed   animation, acid catalyzed

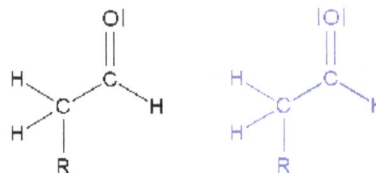

Condensation Types

It is important to distinguish the aldol condensation from other addition reactions of carbonyl compounds.

- When the base is an amine and the active hydrogen compound is sufficiently activated the reaction is called a Knoevenagel condensation.

- In a Perkin reaction the aldehyde is aromatic and the enolate generated from an anhydride.

- A Claisen condensation involves two ester compounds.

- A Dieckmann condensation involves two ester groups in the *same molecule* and yields a cyclic molecule

- A Henry reaction involves an aldehyde and an aliphatic nitro compound.

- A Robinson annulation involves a α,β-unsaturated ketone and a carbonyl group, which first engage in a Michael reaction prior to the aldol condensation.

- In the Guerbet reaction, an aldehyde, formed *in situ* from an alcohol, self-condenses to the dimerized alcohol.

- In the Japp–Maitland condensation water is removed not by an elimination reaction but by a nucleophilic displacement

Aldox Process

In industry the Aldox process developed by Royal Dutch Shell and Exxon, converts propylene and syngas directly to 2-ethylhexanol via hydroformylation to butyraldehyde, aldol condensation to 2-ethylhexenal and finally hydrogenation.

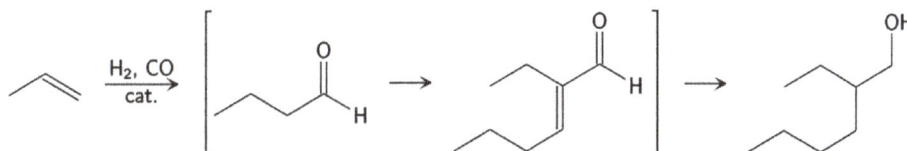

In one study crotonaldehyde is directly converted to 2-ethylhexanal in a palladium / Amberlyst / supercritical carbon dioxide system:

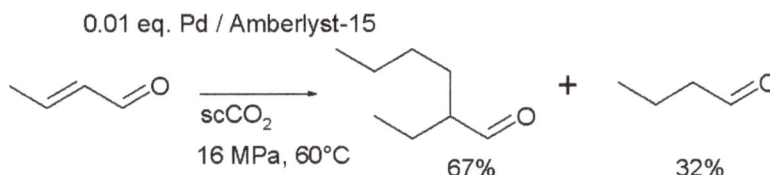

Scope

Ethyl 2-methylacetoacetate and campholenic aldehyde react in an Aldol condensation. The synthetic procedure is typical for this type of reaction. In the process, in addition to water, an equivalent of ethanol and carbon dioxide are lost in decarboxylation.

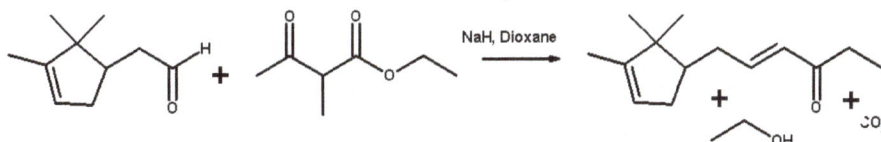

Ethyl glyoxylate **2** and diethyl 2-methylglutaconate **1** react to *isoprenetricarboxylic acid* **3** (isoprene skeleton) with sodium ethoxide. This reaction product is very unstable with initial loss of carbon dioxide and followed by many secondary reactions. This is believed to be due to steric strain resulting from the methyl group and the carboxylic group in the *cis*-dienoid structure.

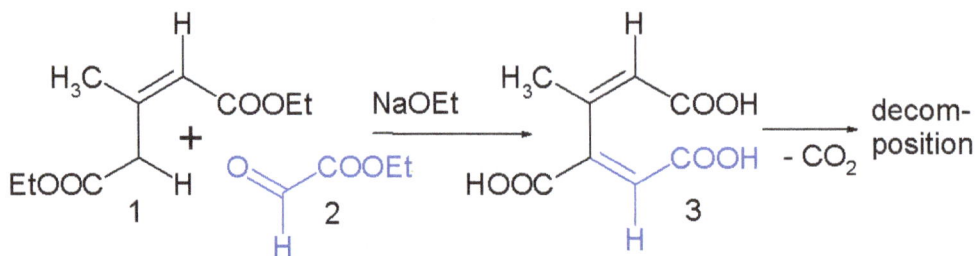

Occasionally an aldol condensation is buried in a multistep reaction or in catalytic cycle such as the one sketched below:

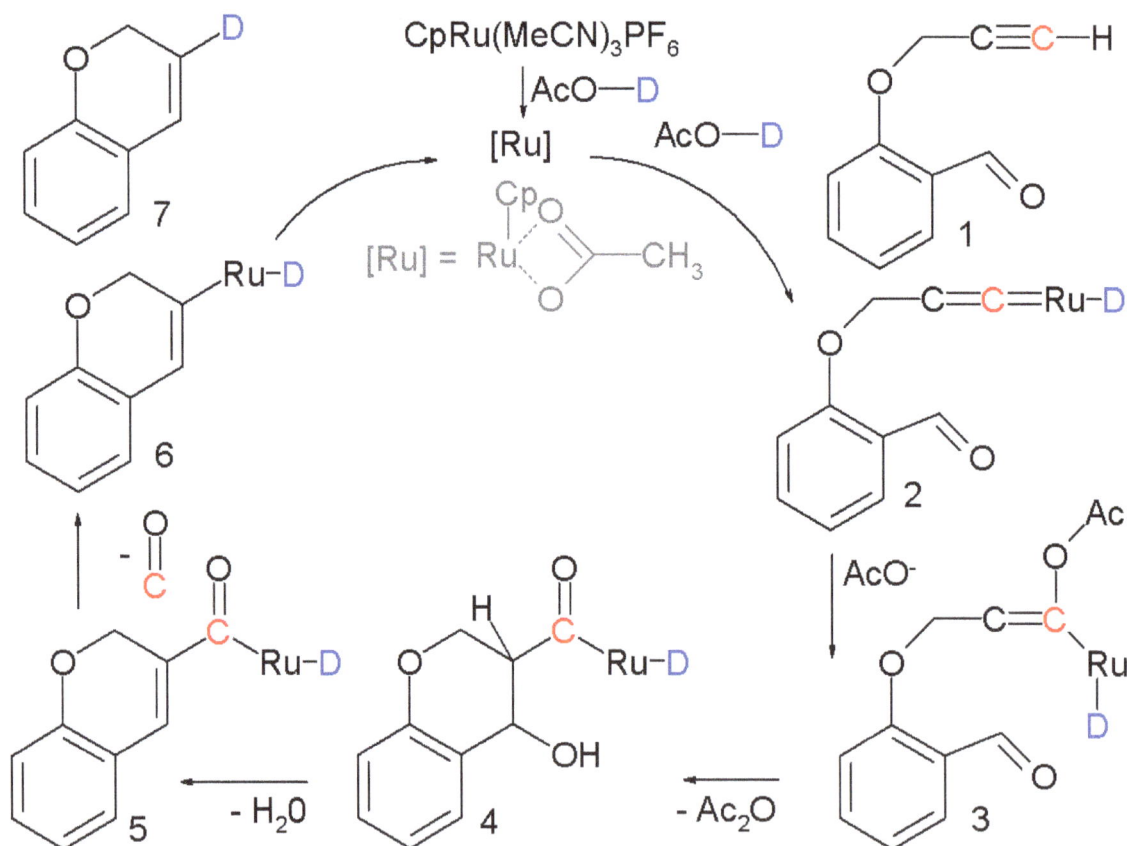

In this reaction an *alkynal* **1** is converted into a cycloalkene **7** with a ruthenium catalyst and the actual condensation takes place with intermediate **3** through **5**. Support for the reaction mechanism is based on isotope labeling.

The reaction between menthone and anisaldehyde is complicated due to steric shielding of the ketone group. This obstacle is overcome by using a strong base such as potassium hydroxide and a very polar solvent such as DMSO in the reaction below:

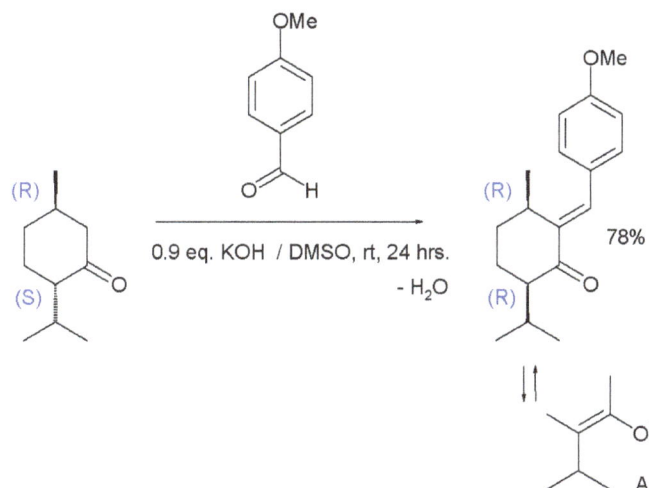

Due to epimerization through a common enolate ion (intermediate **A**) the reaction product has (R,R)-*cis*-configuration and not (R,S)-*trans*-configuration as in the starting material. Because it is only the *cis* isomer that precipitates from solution, this product is formed exclusively.

Arndt–Eistert Reaction

The Arndt–Eistert synthesis is a series of chemical reactions designed to convert a carboxylic acid to a higher carboxylic acid homologue (i.e. contains one additional carbon atom) and is considered a homologation process. Named for the German chemists Fritz Arndt (1885–1969) and Bernd Eistert (1902–1978), Arndt–Eistert synthesis is a popular method of producing β-amino acids from α-amino acids. Acid chlorides react with diazomethane to give diazoketones. In the presence of a nucleophile (water) and a metal catalyst (Ag_2O), diazoketones will form the desired acid homologue.

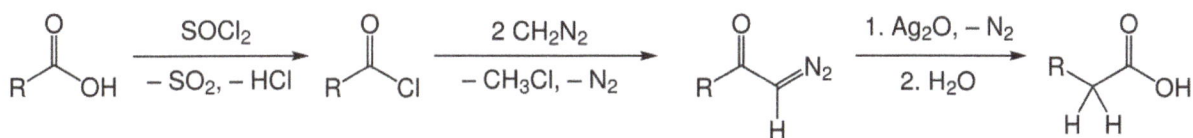

While the classic Arndt–Eistert synthesis uses thionyl chloride to convert the starting acid to an acid chloride, any procedure can be used that will generate an acid chloride.

Diazoketones are typically generated as described here, but other methods such as diazo-group transfer can also apply.

Since diazomethane is toxic and violently explosive, many safer alternatives have been developed, such as the usage of ynolates (Kowalski ester homologation) or diazo(trimethylsilyl)methane.

Reaction Mechanism

The key step in the Arndt–Eistert synthesis is the metal-catalyzed Wolff rearrangement of the diazoketone to form a ketene. In the insertion homologation of t-BOC protected (S)-phenylalanine (2-amino-3-phenylpropanoic acid), *t*-BOC protected (*S*)-3-amino-4-phenylbutanoic acid is formed.

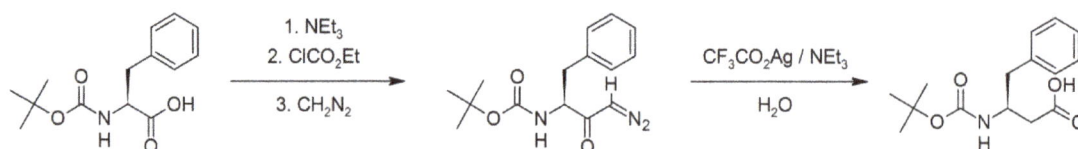

Wolff rearrangement of the α-diazoketone intermediate forms a ketene via a 1,2-rearrangement, which is subsequently hydrolysed to form the carboxylic acid. The consequence of the 1,2-rearrangement is that the methylene group alpha to the carboxyl group in the product is the methylene group from the diazomethane reagant. It has been demonstrated that the rearrangement preserves the stereochemistry of the chiral centre as the product formed from *t*-BOC protected (*S*)-phenylalanine retains the (*S*) stereochemistry with a reported enantiomeric excess of at least 99%.

Heat, light, platinum, silver, and copper salts will also catalyze the Wolff rearrangement to produce the desired acid homologue.

Variations

In the Newman–Beal modification, addition of triethylamine to the diazomethane solution will avoid the formation of α-chloromethylketone side-products.

Baeyer–Villiger Oxidation

The Baeyer-Villiger oxidation (also called Baeyer-Villiger rearrangement) is an organic reaction that forms an ester from a ketone or a lactone from a cyclic ketone. Peroxyacids or peroxides are used as the oxidant. The reaction is named after Adolf Baeyer and Victor Villiger who first reported the reaction in 1899.

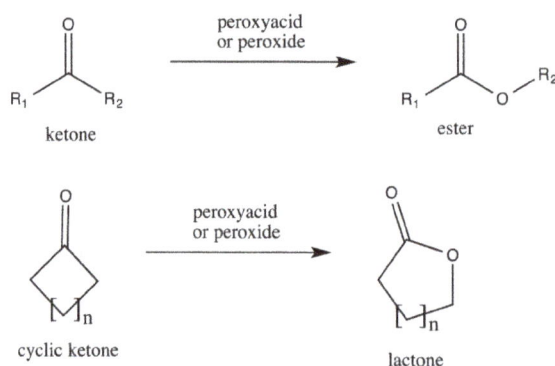

Baeyer-Villiger oxidation

Reaction Mechanism

In the first step of the reaction mechanism, the peroxyacid protonates the oxygen of the carbonyl group. This makes the carbonyl group more susceptible to attack by the peroxyacid. In the next step of the reaction mechanism, the peroxyacid attacks the carbon of the carbonyl group forming what is known as the Criegee intermediate. Through a concerted mechanism, one of the substituents on the ketone migrates to the oxygen of the peroxide group while a carboxylic acid leaves. This migration step is thought to be the rate determining step. Finally, deprotonation of the oxygen of the carbonyl group produces the ester.

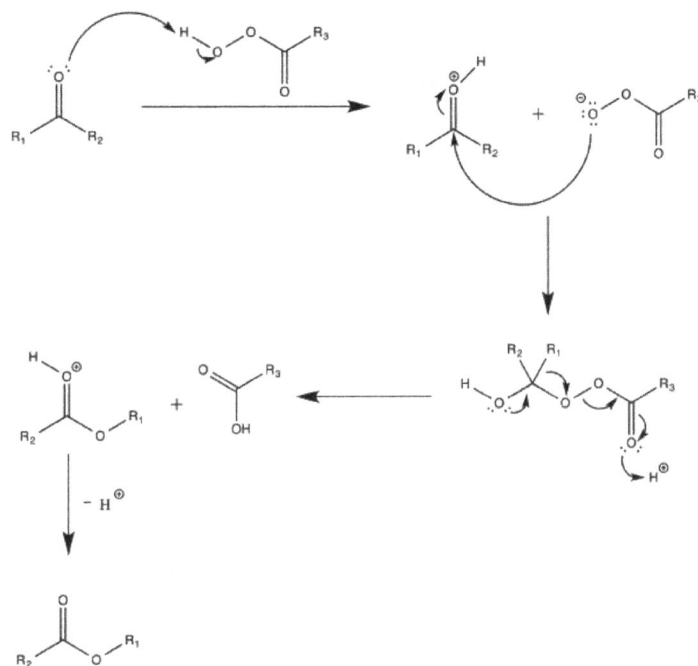

Reaction mechanism of the Baeyer-Villiger oxidation

The products of the Baeyer-Villiger oxidation are believed to be controlled through both primary and secondary stereoelectronic effects. The primary stereoelectronic effect in the Baeyer-Villiger oxidation refers to the necessity of the oxygen-oxygen bond in the peroxide group to be antiperiplanar to the group that migrates. This orientation facilitates optimum overlap of the σ orbital of the migrating group to the σ^* orbital of the peroxide group. The secondary stereoelectronic effect refers to the necessity of the lone pair on the oxygen of the hydroxyl group to be antiperiplanar to the migrating group. This allows for optimum overlap of the oxygen nonbonding orbital with the σ^* orbital of the migrating group. This migration step is also (at least in silico) assisted by two or three peroxyacid units enabling the hydroxyl proton to shuttle to its new position.

R_M = Migrating Group

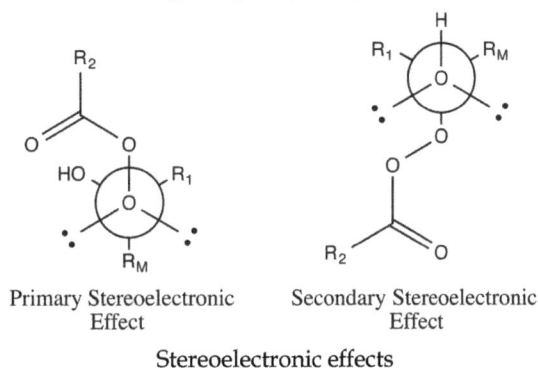

Primary Stereoelectronic   Secondary Stereoelectronic
Effect   Effect

Stereoelectronic effects

The migratory ability is ranked tertiary > secondary > phenyl > primary. Allylic groups also migrate better than primary groups but not as well as secondary groups. If there is an electron withdrawing group on the substituent, then it decreases the rate of migration. There are two explanations for this trend in migration ability. One explanation relies on the carbocation resonance structure of the Criegee intermediate. Keeping this structure in mind, it makes sense that the substituent that can maintain positive charge the best would be most likely to migrate. Tertiary groups are more stable carbocations than secondary groups, and secondary groups are more stable than primary. Therefore, the tertiary > secondary > primary trend is observed.

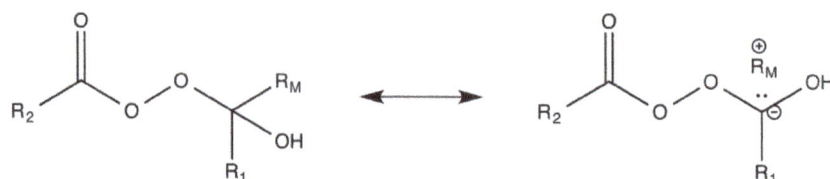

R_M = Migrating Group

Resonance structures of the Criegee intermediate

Another explanation uses stereoelectronic effects and steric bulk to explain the trend. As mentioned, the substituent that is antiperiplanar to the peroxide group in the transition state will be the group that migrates. This transition state has a gauche interaction between the peroxyacid and the non-migrating substituent. If the bulkier group is placed antiperiplanar to the peroxide group, the gauche interaction between the substituent on the forming ester and the carbonyl group of the peroxyacid will be reduced. Thus, it is the bulkier group that ends up antiperiplanar to the peroxide group making it the group that migrates. This explains the trend of tertiary > secondary > primary because tertiary groups are generally bulkier than secondary and primary groups.

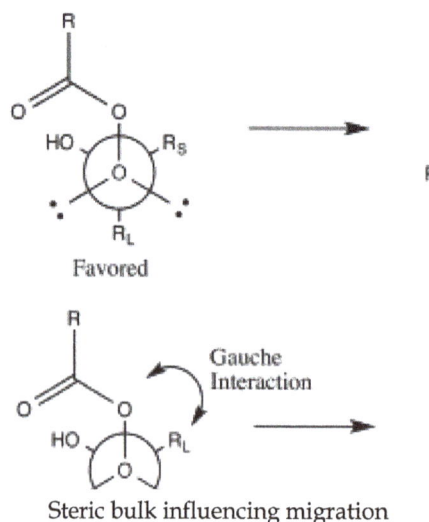

Favored

Steric bulk influencing migration

Historical Background

In 1899, Adolf Baeyer and Victor Villiger first published a demonstration of the reaction that we now know as the Baeyer-Villiger oxidation. They used peroxymonosulfuric acid to make the corresponding lactones from camphor, menthone, and tetrahydrocarvone.

Camphor

Tetrahydrocarvone

Menthone

Original reactions reported by Baeyer and Villiger

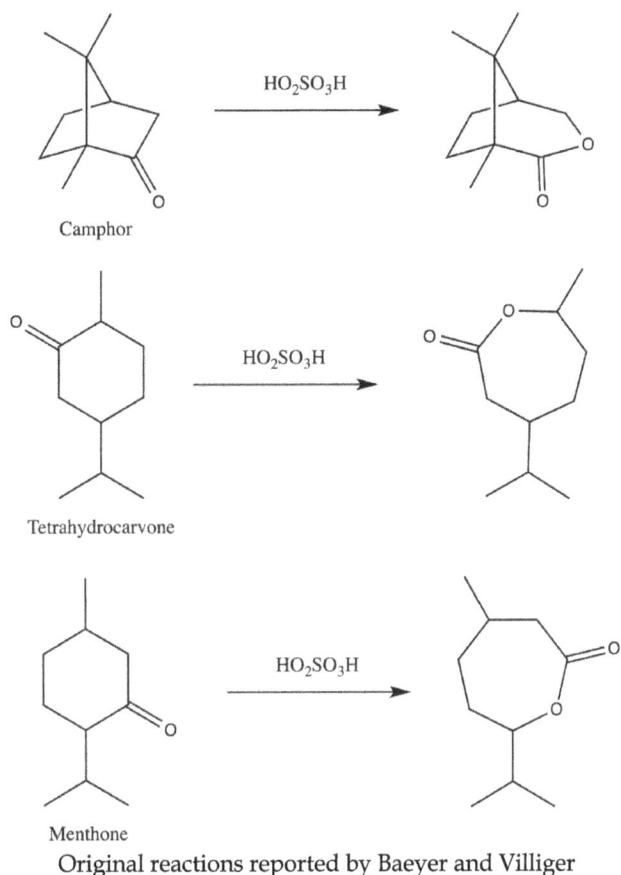

There were three suggested reaction mechanisms of the Baeyer-Villiger oxidation that seemed to fit with observed reaction outcomes. These three reaction mechanisms can really be split into two pathways of peroxyacid attack. The first pathway has the peroxyacid attack the oxygen of the carbonyl group. The second pathway has the peroxyacid attack the carbon of the carbonyl group. The first pathway could lead to two possible intermediates: Baeyer and Villiger suggested a dioxirane intermediate, while Georg Wittig and Gustav Pieper suggested a peroxide intermediate with no dioxirane formation. A second pathway was suggested by Rudolf Criegee. In this pathway, the peracid attacks the carbonyl carbon producing what is now known as the Criegee intermediate.

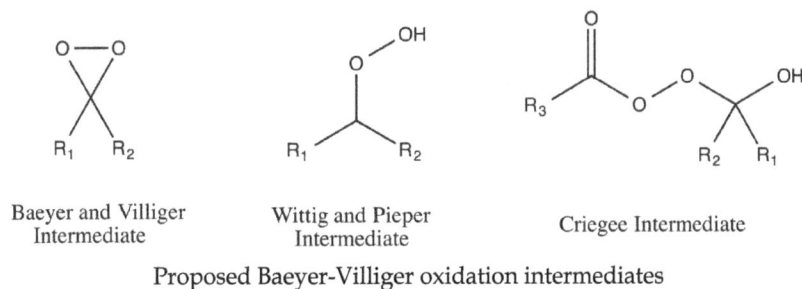

Baeyer and Villiger
Intermediate

Wittig and Pieper
Intermediate

Criegee Intermediate

Proposed Baeyer-Villiger oxidation intermediates

In 1953, William von Eggers Doering and Edwin Dorfman elucidated the correct pathway for the reaction mechanism of the Baeyer-Villiger oxidation by using oxygen-18 to label benzophenone. The three different mechanisms each lead to a different distribution of labelled products. The Criegee intermediate leads to a product that is only labelled on the oxygen of the carbonyl group.

The product of the Wittig and Pieper intermediate is only labeled on the oxygen of the ester. The Baeyer and Villiger intermediate leads to a 1:1 distribution of both of the above products. The outcome of the labelling experiment supported the Criegee intermediate. It is now believed that the mechanism follows the Criegee intermediate.

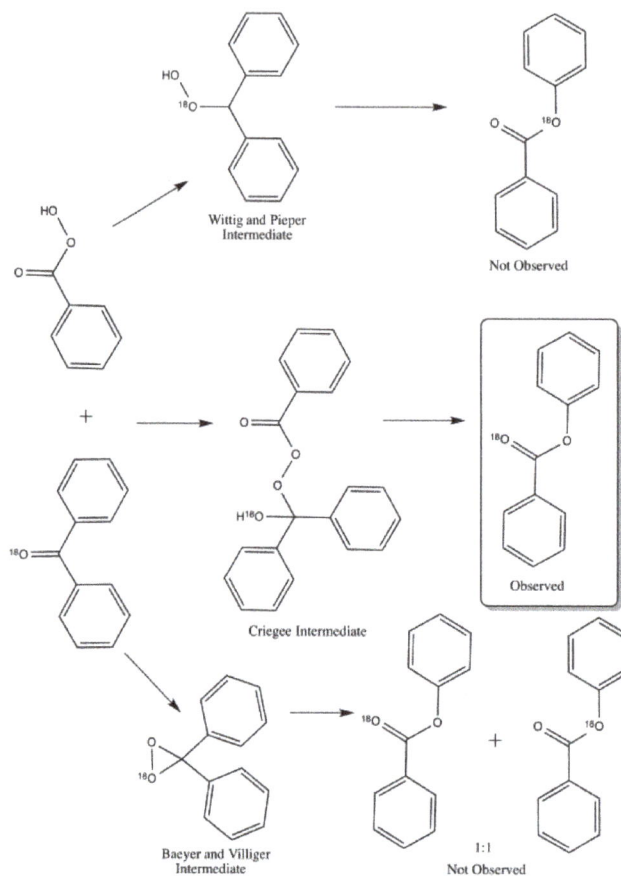

The different possible outcomes of Dorfman and Doering's labelling experiment

Stereochemistry

The migration does not change the stereochemistry of the group that transfers. Therefore, if it is a chiral group that migrates, the chirality of that group will not be changed.

Reagents

Although many different peroxyacids are used for the Baeyer-Villiger oxidation, some of the more common oxidants include *meta*-chloroperbenzoic acid (mCPBA) and trifluoroperacetic acid (TFPAA). The reactivity differs depending on the choice of the peroxyacid. The general trend of reactivity correlates to the strength of the corresponding acid (or alcohol in the case of the peroxides). The stronger the acid, the more reactive will the corresponding peroxyacid be in performing the Baeyer-Villiger oxidation. The trend of reactivity of some reagents is TFPAA > 4-nitroperbenzoic acid > mCPBA and performic acid > peracetic acid > hydrogen peroxide > tert-butyl hydroperoxide. The peroxides are much less reactive than the peroxyacids. In fact, hydrogen peroxide requires a catalyst in order to be used as an oxidant in the Baeyer-Villiger oxidation.

Limitations

The use of peroxyacids and peroxides when performing the Baeyer-Villiger oxidation can cause the undesirable oxidation of other functional groups. Alkenes and amines are a few of the groups that can be oxidized. However, methods have been developed that will allow for the tolerance of these functional groups. For instance, if there is an alkene present in the ketone, the alkene could potentially undergo oxidation to the epoxide. In general, electron-poor alkenes will prefer the Baeyer-Villiger oxidation, while electron-rich will prefer the epoxidation. However, it may depend on the reagents that are used. For example, there are methods that will selectively choose the formation of the epoxide or the ester. In 1962, G. B. Payne reported that the use of hydrogen peroxide in the presence of a selenium catalyst will produce the epoxide, while use of peroxyacetic acid will form the ester.

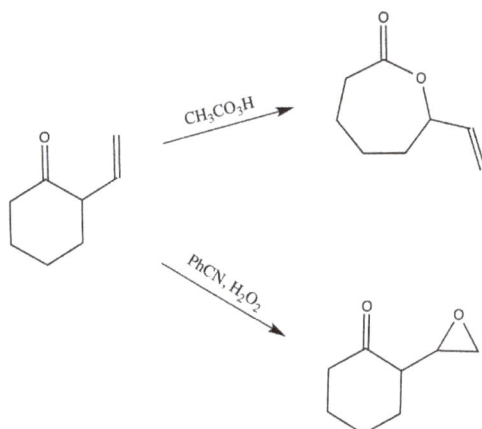

Payne reported that different reagents will give different outcomes when there are more than one functional group

Modifications

Catalytic Baeyer-Villiger oxidation

There has been interest in making the Baeyer-Villiger oxidation work with hydrogen peroxide as an oxidant in the presence of a catalyst. Using hydrogen peroxide as an oxidant makes the reaction more environmentally friendly as the waste produced would just be water. The use of benzeneseleninic acid derivatives as a catalyst has been reported to give high selectivity with hydrogen peroxide as the oxidant.

Baeyer-Villiger Monooxygenases

Another way to create a catalytic Baeyer-Villiger oxidation is by using enzymes as the catalyst. Baeyer-Villiger monooxygenases (BVMOs) use dioxygen to perform the Baeyer-Villiger oxidation. These enzymes are capable of enantioselective oxidations of prochiral substrates.

Asymmetric Baeyer-Villiger oxidation

There have been attempts to use organometallic catalysts to perform an enantioselective Baeyer-Villiger oxidation. The first reported instance of an asymmetric Baeyer-Villiger oxidation on a prochiral ketone used dioxygen as the oxidant and a copper catalyst. Other catalysts followed such as platinum and aluminum catalysts.

Applications

Zoapatanol

Zoapatanol is a biologically active molecule that occurs naturally in the zeopatle plant. The zeopatle plant has been used in Mexico to make a tea that can induce menstruation and labor. In 1981, Vinayak Kane and Donald Doyle reported a synthesis of zoapatanol. They used the Baeyer-Villiger oxidation to make a lactone that served as a crucial building block that ultimately led to the synthesis of zoapatanol.

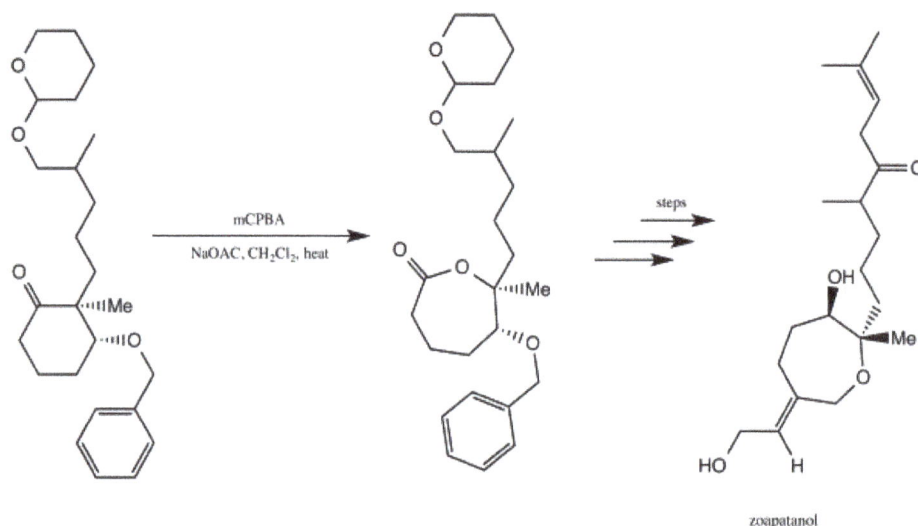

Kane and Doyle used a Baeyer-Villiger oxidation to synthesize zoapatanol

Steroids

Steroids are an important class of molecules for use in therapeutics. For instance, testololactone has been identified as an anticancer agent. In 2013, Alina Świzdor reported the transformation of dehydroepiandrosterone to testololactone by use of a fungus that produces Baeyer-Villiger monooxygenases. The fungus formed testololactone from dehydroepiandrosterone via a Baeyer-Villiger oxidation.

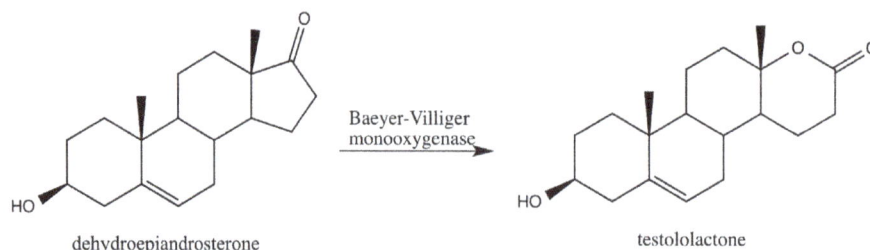

Świzdor reported that a Baeyer-Villiger monooxygenase could change dehydroepiandrosterone into testololactone

Dakin Oxidation

The Dakin oxidation (or Dakin reaction) is an organic redox reaction in which an *ortho*- or *para*-hydroxylated phenyl aldehyde (2-hydroxybenzaldehyde or 4-hydroxybenzaldehyde) or ketone reacts with hydrogen peroxide in base to form a benzenediol and a carboxylate. Overall, the carbonyl group is oxidized, and the hydrogen peroxide is reduced.

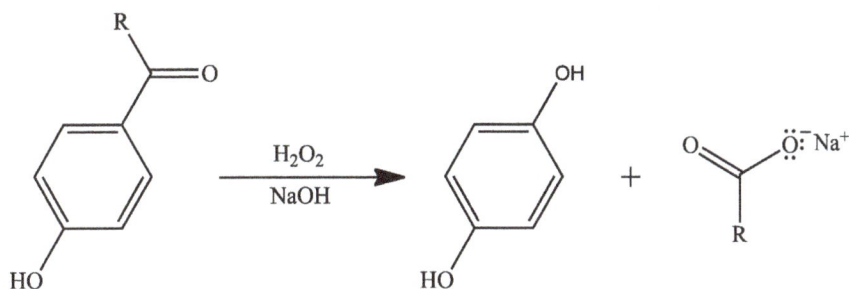

The Dakin oxidation

The Dakin oxidation, which is closely related to the Baeyer-Villiger oxidation, is not to be confused with the Dakin-West reaction, though both are named after Henry Drysdale Dakin.

Reaction Mechanism

The Dakin oxidation starts with nucleophilic addition of a hydroperoxide anion to the carbonyl carbon, forming a tetrahedral intermediate (**2**). The intermediate collapses, causing [1,2]-aryl migration, hydroxide elimination, and formation of a phenyl ester (**3**). The phenyl ester is subsequently hydrolyzed: nucleophilic addition of hydroxide from solution to the ester carbonyl carbon forms a second tetrahedral intermediate (**4**), which collapses, eliminating a phenoxide and forming a carboxylic acid (**5**). Finally, the phenoxide extracts the acidic hydrogen from the carboxylic acid, yielding the collected products (**6**).

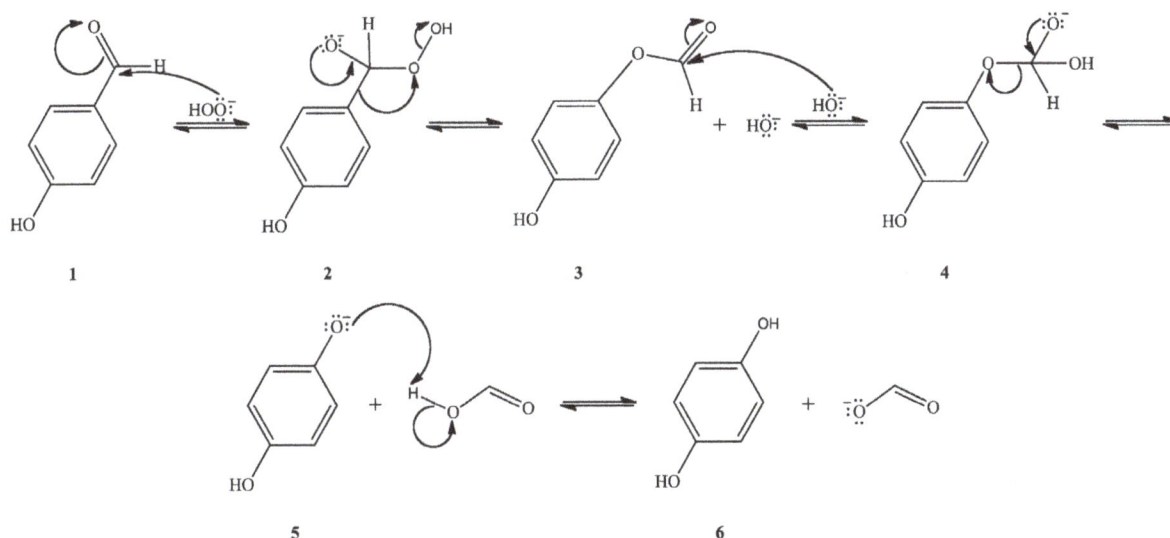

Base-catalyzed Dakin oxidation mechanism

Factors Affecting Reaction Kinetics

The Dakin oxidation has two rate-limiting steps: nucleophilic addition of hydroperoxide to the carbonyl carbon and [1,2]-aryl migration. Therefore, the overall rate of oxidation is dependent on the nucleophilicity of hydroperoxide, the electrophilicity of the carbonyl carbon, and the speed of

[1,2]-aryl migration. The alkyl substituents on the carbonyl carbon, the relative positions of the hydroxyl and carbonyl groups on the aryl ring, the presence of other functional groups on the ring, and the reaction mixture pH are four factors that affect these rate-limiting steps.

Alkyl Substituents

In general, phenyl aldehydes are more reactive than phenyl ketones because the ketone carbonyl carbon is less electrophilic than the aldehyde carbonyl carbon. The difference can be mitigated by increasing the temperature of the reaction mixture.

Relative Positions of Hydroxyl and Carbonyl Groups

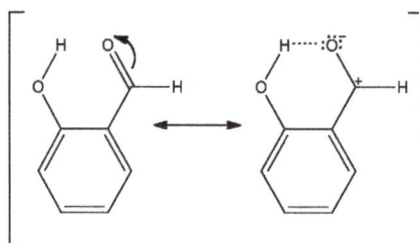

7

Hydrogen bond in *ortho* substrate

O-hydroxy phenyl aldehydes and ketones oxidize faster than *p*-hydroxy phenyl aldehydes and ketones in weakly basic conditions. In *o*-hydroxy compounds, when the hydroxyl group is protonated, an intramolecular hydrogen bond can form between the hydroxyl hydrogen and the carbonyl oxygen, stabilizing a resonance structure with positive charge on the carbonyl carbon, thus increasing the carbonyl carbon's electrophilicity (7). Lacking this stabilization, the carbonyl carbon of *p*-hydroxy compounds is less electrophilic. Therefore, *o*-hydroxy compounds are oxidized faster than *p*-hydroxy compounds when the hydroxyl group is protonated.

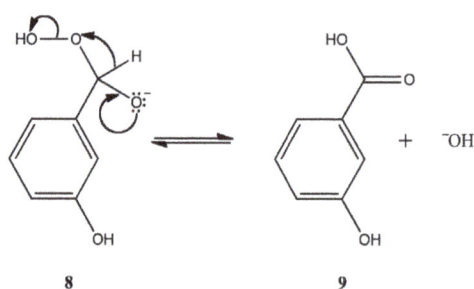

8   **9**

Carboxylic acid product formation

M-hydroxy compounds do not oxidize to *m*-benzenediols and carboxylates. Rather, they form phenyl carboxylic acids. Variations in the aryl rings' migratory aptitudes can explain this. Hydroxyl groups *ortho* or *para* to the carbonyl group concentrate electron density at the aryl carbon bonded to the carbonyl carbon (10c, 11d). Phenyl groups have low migratory aptitude, but higher electron density at the migrating carbon increases migratory aptitude, facilitating [1,2]-aryl migration and allowing the reaction to continue. *M*-hydroxy compounds do not concentrate electron density at the migrating carbon (12a, 12b, 12c, 12d); their aryl groups' migratory aptitude remains low. The benzylic hydrogen, which has the highest migratory aptitude, migrates instead (8), forming a phenyl carboxylic acid (9).

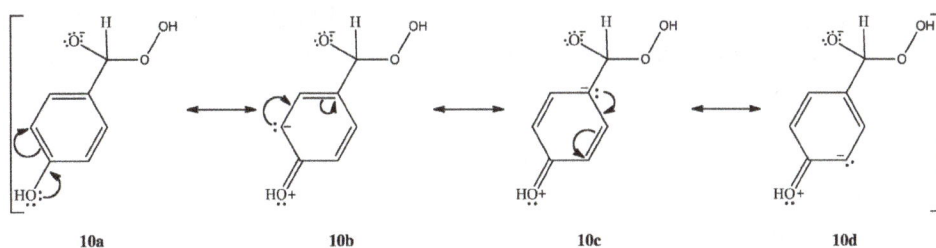

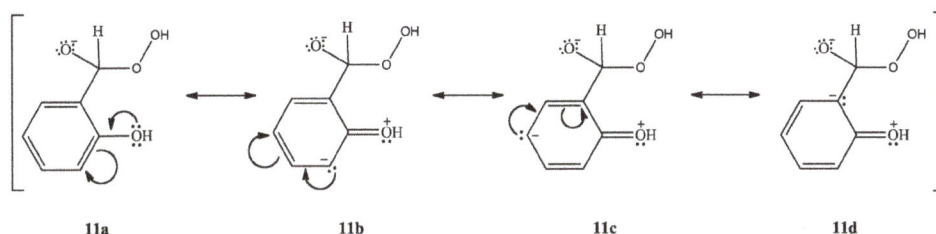

Concentration of electron density at the migrating carbon with *para* (top) and *ortho* (bottom) hydroxyl group

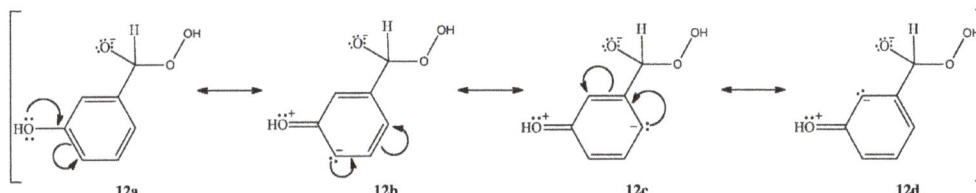

Lack of electron density concentration at the migrating carbon with *meta* hydroxyl group

Other Functional Groups on the Aryl ring

Substitution of phenyl hydrogens with electron-donating groups *ortho* or *para* to the carbonyl group increases electron density at the migrating carbon, promotes [1,2]-aryl migration, and accelerates oxidation. Substitution with electron-donating groups *meta* to the carbonyl group does not change electron density at the migrating carbon; because unsubstituted phenyl group migratory aptitude is low, hydrogen migration dominates. Substitution with electron-withdrawing groups *ortho* or *para* to the carbonyl decreases electron density at the migrating carbon (**13c**), inhibits [1,2]-aryl migration, and favors hydrogen migration.

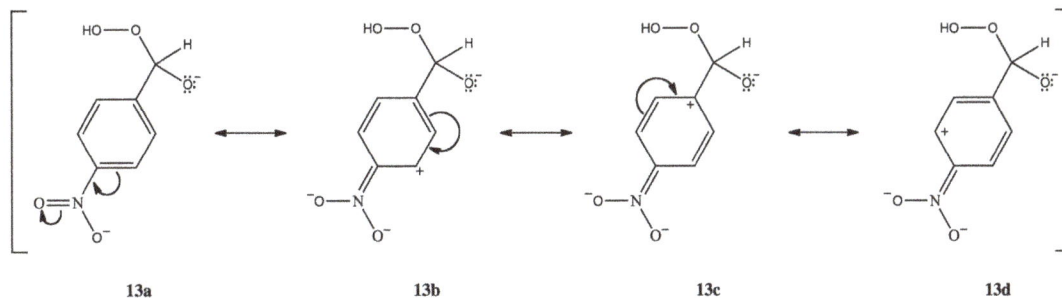

Concentration of positive charge at migrating carbon with *para* nitro group

pH

The hydroperoxide anion is a more reactive nucleophile than neutral hydrogen peroxide. Consequently, oxidation accelerates as pH increases toward the pK_a of hydrogen peroxide and hydroperoxide concentration climbs. At pH higher than 13.5, however, oxidation does not occur, possibly due to deprotonation of the second peroxidic oxygen. Deprotonation of the second peroxidic oxygen would prevent [1,2]-aryl migration because the lone oxide anion is too basic to be eliminated (2).

Deprotonation of the hydroxyl group increases electron donation from the hydroxyl oxygen. When the hydroxyl group is *ortho* or *para* to the carbonyl group, deprotonation increases the electron density at the migrating carbon, promoting faster [1,2]-aryl migration. Therefore, [1,2]-aryl migration is facilitated by the pH range that favors deprotonated over protonated hydroxyl group.

Variants

Acid-catalyzed Dakin Oxidation

The Dakin oxidation can occur in mild acidic conditions as well, with a mechanism analogous to the base-catalyzed mechanism. In methanol, hydrogen peroxide, and catalytic sulfuric acid, the carbonyl oxygen is protonated (14), after which hydrogen peroxide adds as a nucleophile to the carbonyl carbon, forming a tetrahedral intermediate (15). Following an intramolecular proton transfer (16,17), the tetrahedral intermediate collapses, [1,2]-aryl migration occurs, and water is eliminated (18). Nucleophilic addition of methanol to the carbonyl carbon forms another tetrahedral intermediate (19). Following a second intramolecular proton transfer (20,21), the tetrahedral intermediate collapses, eliminating a phenol and forming an ester protonated at the carbonyl oxygen (22). Finally, deprotonation of the carbonyl oxygen yields the collected products and regenerates the acid catalyst (23).

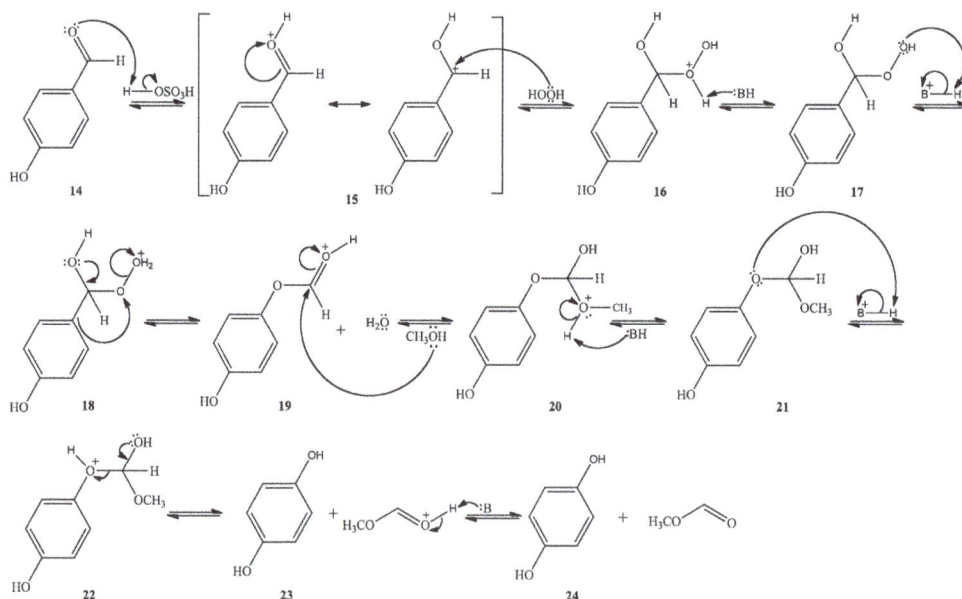

Acid-catalyzed Dakin oxidation mechanism

Boric acid-catalyzed Dakin Oxidation

Adding boric acid to the acid-catalyzed reaction mixture increases the yield of phenol product over phenyl carboxylic acid product, even when using phenyl aldehyde or ketone reactants with electron-donating groups *meta* to the carbonyl group or electron-withdrawing groups *ortho* or *para* to the carbonyl group. Boric acid and hydrogen peroxide form a complex in solution that, once added to the carbonyl carbon, favors aryl migration over hydrogen migration, maximizing the yield of phenol and reducing the yield of phenyl carboxylic acid.

Methyltrioxorhenium-catalyzed Dakin Oxidation

Using an ionic liquid solvent with catalytic methyltrioxorhenium (MTO) dramatically accelerates Dakin oxidation. MTO forms a complex with hydrogen peroxide that increases the rate of addition of hydrogen peroxide to the carbonyl carbon. MTO does not, however, change the relative yields of phenol and phenyl carboxylic acid products.

Urea-catalyzed Dakin Oxidation

Mixing urea and hydrogen peroxide yields urea-hydrogen peroxide complex (UHC). Adding dry UHC to solventless phenyl aldehyde or ketone also accelerates Dakin oxidation. Like MTO, UHP increases the rate of nucleophilic addition of hydrogen peroxide. But unlike the MTO-catalyzed variant, the urea-catalyzed variant does not produce potentially toxic heavy metal waste.

Synthetic Applications

The Dakin oxidation is most commonly used to synthesize benzenediols and alkoxyphenols. Catechol, for example, is synthesized from o-hydroxy and o-alkoxy phenyl aldehydes and ketones, and is used as the starting material for synthesis of several compounds, including the catecholamines, catecholamine derivatives, and 4-tert-butylcatechol, a common antioxidant and polymerization inhibitor. Other synthetically useful products of the Dakin oxidation include guaiacol, a precursor of several flavorants; hydroquinone, a common photograph-developing agent; and 2-tert-butyl-4-hydroxyanisole and 3-tert-butyl-4-hydroxyanisole, two antioxidants commonly used to preserve packaged food. In addition, the Dakin oxidation is useful in the synthesis of indolequinones, naturally-occurring compounds that exhibit high anti-biotic, anti-fungal, and anti-tumor activities.

Biginelli Reaction

The Biginelli reaction is a multiple-component chemical reaction that creates 3,4-dihydropyrimidin-2(1H)-ones 4 from ethyl acetoacetate 1, an aryl aldehyde (such as benzaldehyde 2), and urea 3. It is named for the Italian chemist Pietro Biginelli.

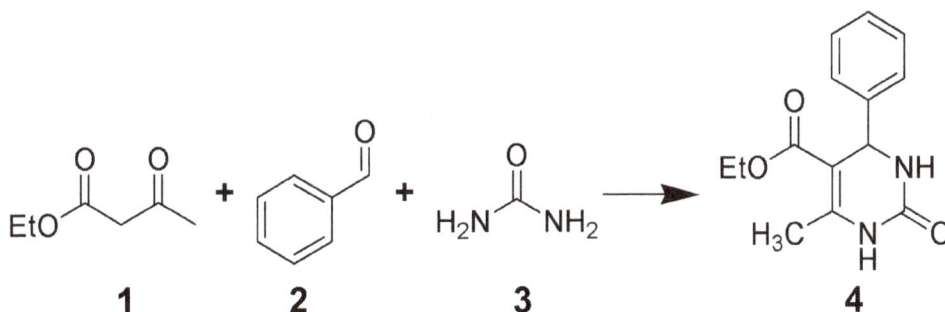

This reaction was developed by Pietro Biginelli in 1891. The reaction can be catalyzed by Brønsted acids and/or by Lewis acids such as copper(II) trifluoroacetate hydrate and boron trifluoride. Several solid-phase protocols utilizing different linker combinations have been published.

Dihydropyrimidinones, the products of the Biginelli reaction, are widely used in the pharmaceutical industry as calcium channel blockers, antihypertensive agents, and alpha-1-a-antagonists.

Reaction Mechanism

The reaction mechanism of the Biginelli reaction is a series of bimolecular reactions leading to the desired dihydropyrimidinone.

According to a mechanism proposed by Sweet in 1973 the aldol condensation of ethylacetoacetate **1** and the aryl aldehyde is the rate-limiting step leading to the carbenium ion **2**. The nucleophilic addition of urea gives the intermediate **4**, which quickly dehydrates to give the desired product **5**.

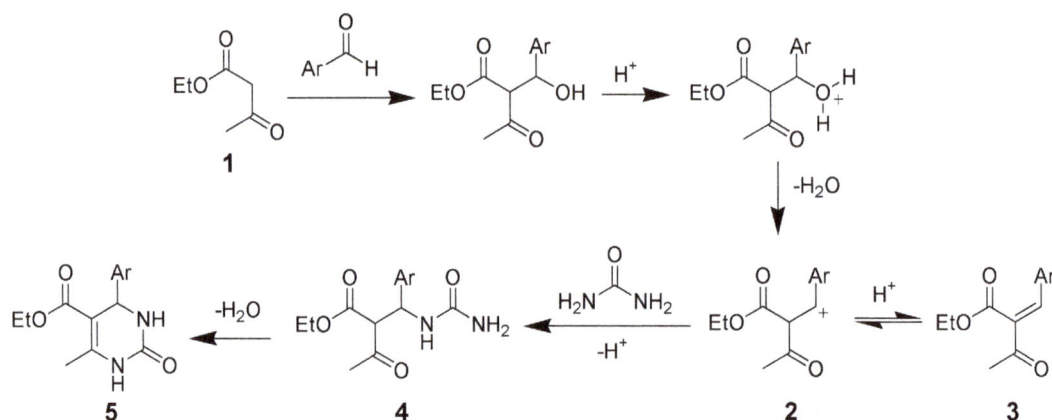

This mechanism is superseded by one by Kappe in 1997:

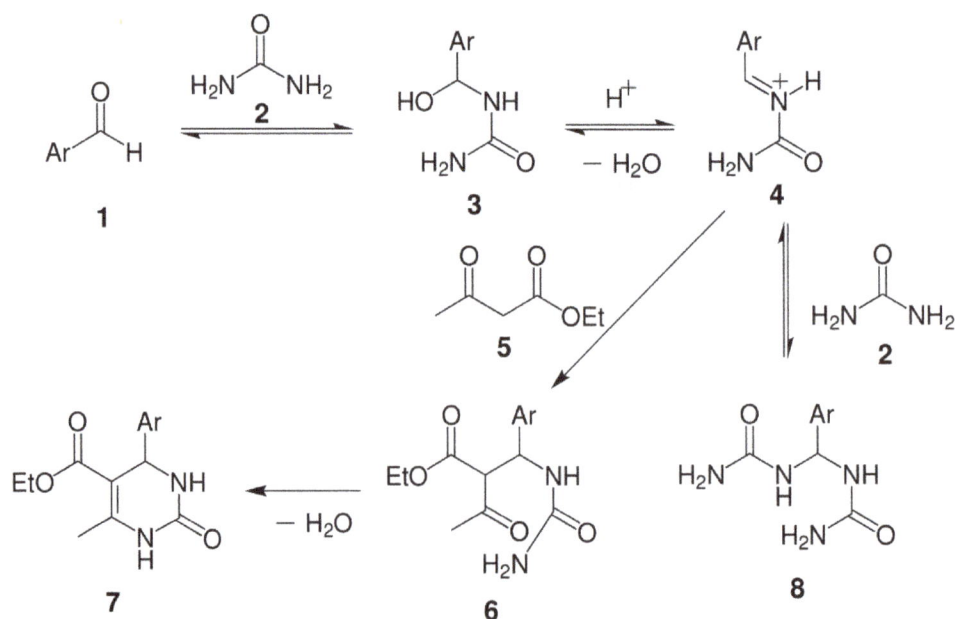

This scheme begins with rate determining nucleophilic addition by the urea to the aldehyde. The ensuing condensation step is catalyzed by the addition of acid, resulting in the imine nitrogen. The β-ketoester then adds to the imine bond and consequently the ring is closed by the nucleophilic attack by the amine onto the carbonyl group. This final step ensues a second condensation and results in the Biginelli compound.

Advances in Biginelli Reaction

In 1987, Atwal *et al.* reported a modification to the Biginelli reaction that consistently generated higher yields. Atul Kumar has reported first enzymatic synthesis for Biginelli reaction via yeast catalysed protocol in high yields.

Simmons–Smith Reaction

The Simmons–Smith reaction is an organic cheletropic reaction involving an organozinc carbenoid that reacts with an alkene (or alkyne) to form a cyclopropane. It is named after Howard Ensign Simmons, Jr. and Ronald D. Smith. It uses a methylene free radical intermediate that is delivered to both carbons of the alkene simultaneously, therefore the configuration of the double bond is preserved in the product and the reaction is stereospecific.

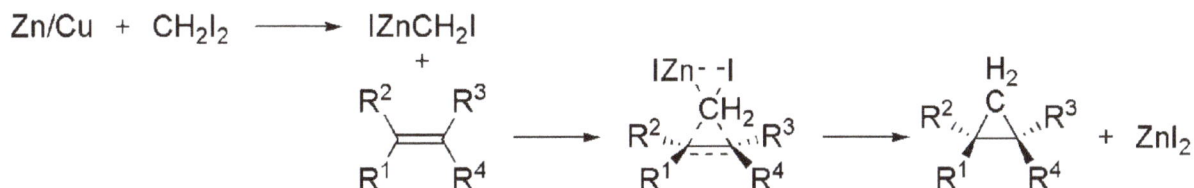

Thus, cyclohexene, diiodomethane, and a zinc-copper couple (as iodomethylzinc iodide, ICH_2ZnI) yield norcarane (bicyclo[4.1.0]heptane).

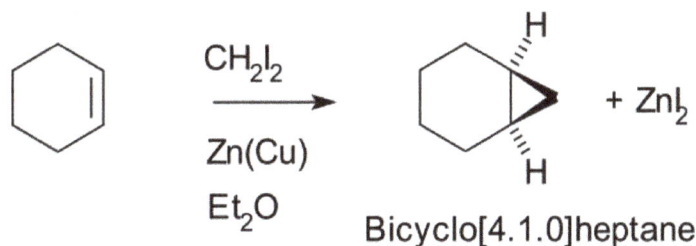

Bicyclo[4.1.0]heptane

The Simmons–Smith reaction is generally preferred over other methods of cyclopropanation, however it can be expensive due to the high cost of diiodomethane. Modifications involving cheaper alternatives have been developed, such as dibromomethane or diazomethane and zinc iodide. The reactivity of the system can also be increased by exchanging the zinccopper couple for diethylzinc, however as this reagent is pyrophoric it must be handled carefully.

The Simmons–Smith reaction is generally subject to steric effects, and thus cyclopropanation usually takes place on the less hindered face. However, when a hydroxy substituent is present in the substrate in proximity to the double bond, the zinc coordinates with the hydroxy substituent, directing cyclopropanation *cis* to the hydroxyl group (which may not correspond to cyclopropanation of the sterically most accessible face of the double bond): An interactive 3D model of this reaction can be seen here (java required).

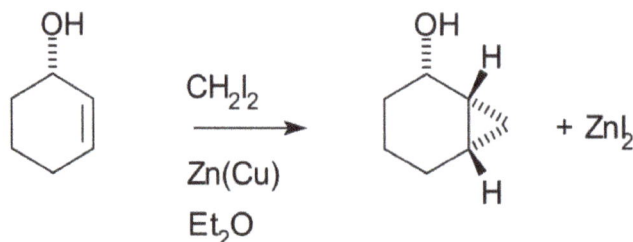

The Simmons–Smith reagent, namely diiodomethane and diethylzinc, can react with allylic thioethers to generate sulfur ylides, which can subsequently undergo a 2,3-sigmatropic rearrangement, and will not cyclopropanate an alkene in the same molecule unless excess Simmons–Smith reagent is used:

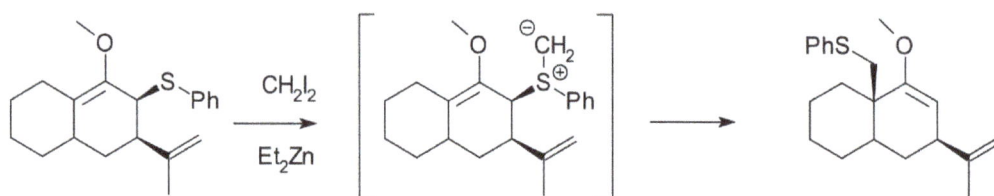

Asymmetric Simmons–Smith reaction

Although asymmetric cyclopropanation methods based on diazo compounds exist since 1966, the asymmetric Simmons–Smith reaction was introduced in 1992 with a reaction of cinnamyl alcohol with diethylzinc, diiodomethane and a chiral disulfonamide in dichloromethane:

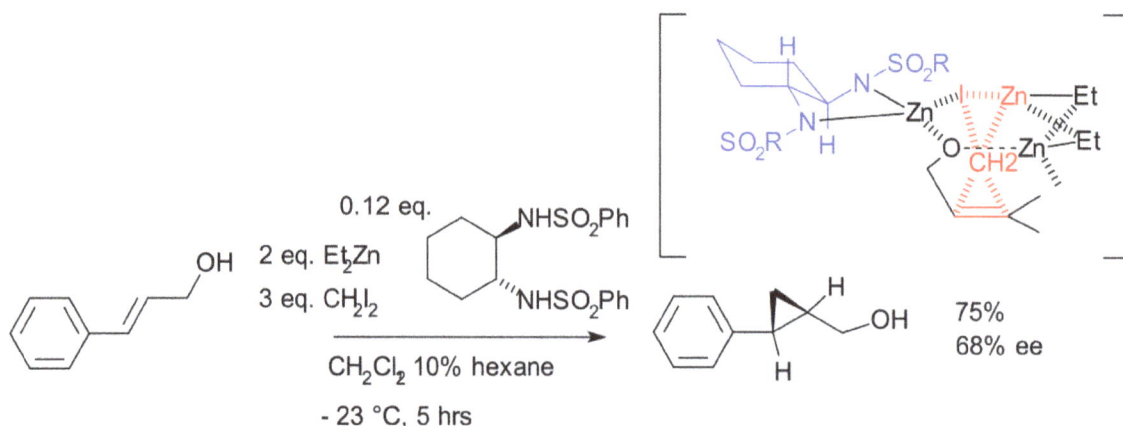

The hydroxyl group is a prerequisite serving as an anchor for zinc. An interactive 3D model of a similar reaction can be seen here (java required). In another version of this reaction the ligand is based on salen and Lewis acid DIBAL is added:

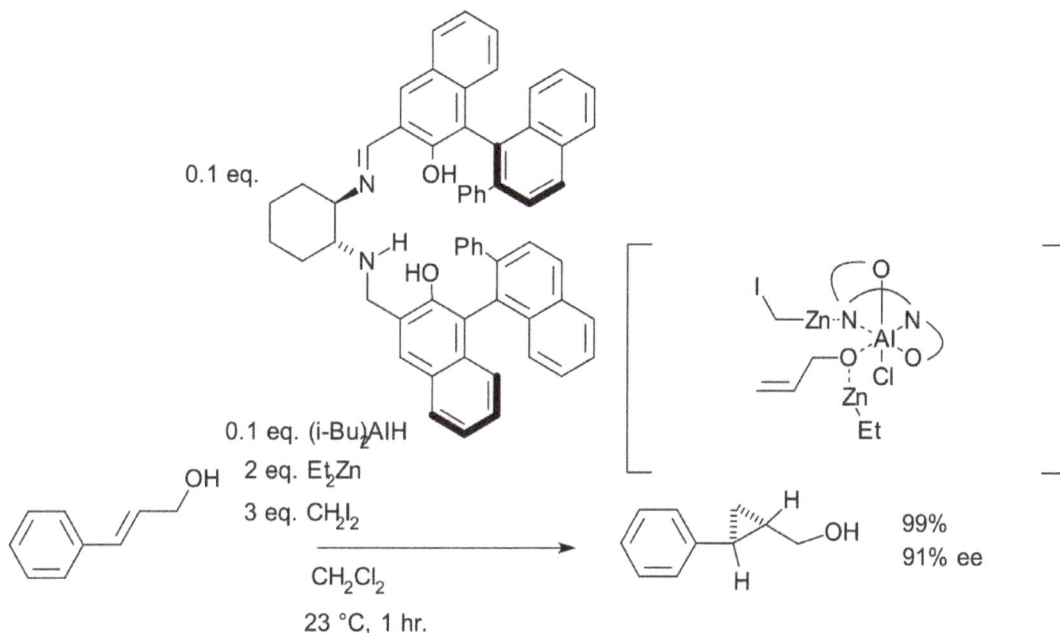

Williamson Ether Synthesis

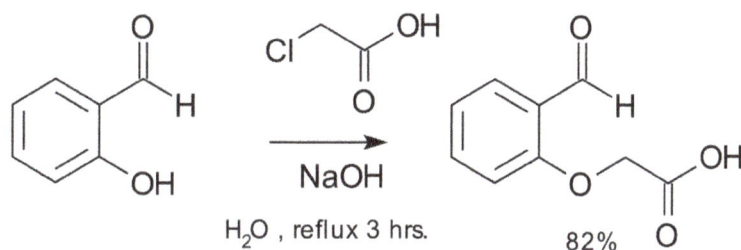

Ether synthesis by reaction of salicylaldehyde with chloroacetic acid and sodium hydroxide

The Williamson ether synthesis is an organic reaction, forming an ether from an organohalide and a deprotonated alcohol (alkoxide). This reaction was developed by Alexander Williamson in 1850. Typically it involves the reaction of an alkoxide ion with a primary alkyl halide via an S_N2 reaction. This reaction is important in the history of organic chemistry because it helped prove the structure of ethers.

The general reaction mechanism is as follows:

An example is the reaction of sodium ethoxide with chloroethane to form diethyl ether and sodium chloride:

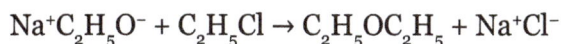

$$Na^+C_2H_5O^- + C_2H_5Cl \rightarrow C_2H_5OC_2H_5 + Na^+Cl^-$$

Scope

The Williamson reaction is of broad scope, is widely used in both laboratory and industrial synthesis, and remains the simplest and most popular method of preparing ethers. Both symmetrical and asymmetrical ethers are easily prepared. The intramolecular reaction of halohydrins in particular, gives epoxides.

In the case of asymmetrical ethers there are two possibilities for the choice of reactants, and one is usually preferable either on the basis of availability or reactivity. The Williamson reaction is also frequently used to prepare an ether indirectly from two alcohols. One of the alcohols is first converted to a leaving group (usually tosylate), then the two are reacted together.

The alkoxide (or aroxide) may be primary, secondary or tertiary. The alkylating agent, on the other hand is most preferably primary. Secondary alkylating agents also react, but tertiary ones are usually too prone to side reactions to be of practical use. The leaving group is most often a halide or a sulfonate ester synthesized for the purpose of the reaction. Since the conditions of the reaction are rather forcing, protecting groups are often used to pacify other parts of the reacting molecules (e.g. other alcohols, amines, etc.)

Conditions

Since alkoxide ions are highly reactive, they are usually prepared immediately prior to the reaction, or are generated *in situ*. In laboratory chemistry, *in situ* generation is most often accomplished by the use of a carbonate base or potassium hydroxide, while in industrial syntheses phase transfer catalysis is very common. A wide range of solvents can be used, but protic solvents and apolar solvents tend to slow the reaction rate strongly, as a result of lowering the availability of the free nucleophile. For this reason, acetonitrile and N,N-dimethylformamide are particularly commonly used.

A typical Williamson reaction is conducted at 50–100 °C and is complete in 1–8 hours. Often the complete disappearance of the starting material is difficult to achieve, and side reactions are common. Yields of 50–95% are generally achieved in laboratory syntheses, while near-quantitative conversion can be achieved in industrial procedures.

Catalysis is not usually necessary in laboratory syntheses. However, if an unreactive alkylating agent is used (e.g. an alkyl chloride) then the rate of reaction can be greatly improved by the addition of a catalytic quantity of a soluble iodide salt (which undergoes halide exchange with the chloride to yield a much more reactive iodide, a variant of the Finkelstein reaction). In extreme cases, silver salts may be added for example silver oxide:

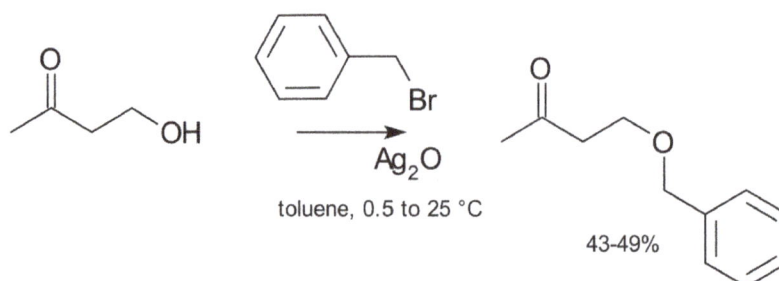

The silver ion coordinates with the halide leaving group to make its departure more facile. Finally, phase transfer catalysts are sometimes used (e.g. tetrabutylammonium bromide or 18-crown-6) in order to increase the solubility of the alkoxide by offering a softer counter-ion.

Side Reactions

The Williamson reaction often competes with the base-catalyzed elimination of the alkylating agent, and the nature of the leaving group as well as the reaction conditions (particularly the temperature and solvent) can have a strong effect on which is favored. In particular, some structures of alkylating agent can be particularly prone to elimination.

When the nucleophile is an aroxide ion, the Williamson reaction can also compete with alkylation on the ring since the aroxide is an ambident nucleophile.

Pechmann Condensation

The Pechmann condensation is a synthesis of coumarins, starting from a phenol and a carboxylic acid or ester containing a β-carbonyl group. The condensation is performed under acidic conditions. The mechanism involves an esterification/transesterification followed by attack of the activated carbonyl ortho to the oxygen to generate the new ring. The final step is a dehydration, as seen following an aldol condensation. It was discovered by the German chemist Hans von Pechmann.

With simple phenols, the conditions are harsh, although yields may still be good.

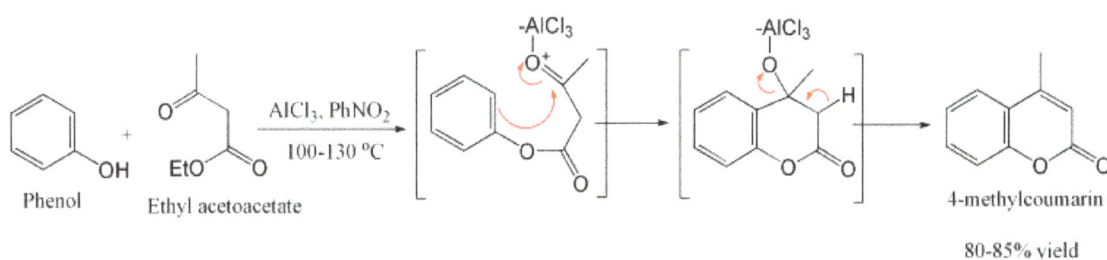

With highly activated phenols such as resorcinol, the reaction can be performed under much milder conditions. This provides a useful route to umbelliferone derivatives:

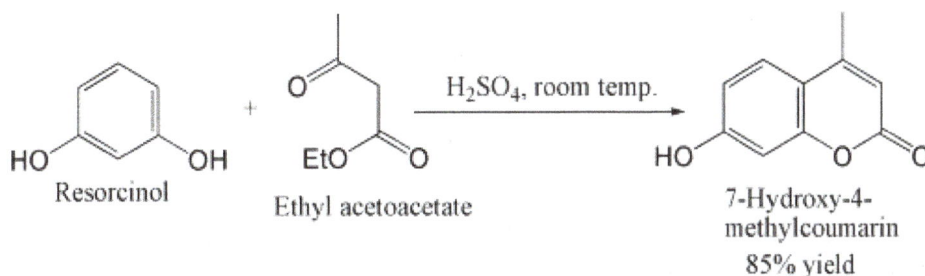

For coumarins unsubstituted at the 4-position, the method requires the use of formylacetic acid or ester. These are unstable and not commercially available, but the acid may be produced *in situ* from malic acid and sulfuric acid above 100 °C. As soon as it forms, the formylacetic acid performs the Pechmann condensation. In the example shown, umbelliferone itself is produced, albeit in low yield:

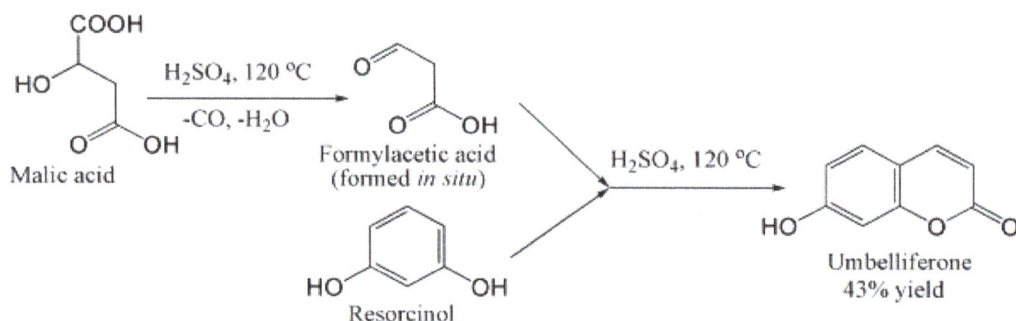

Mechanistic Studies

The mechanism of the reaction has been studied in details with theoretical treatment.

The study shown that reaction takes place on oxo-form, and not on enolic-form. Three different oxo-routes have been proposed.

Simonis Chromone Cyclization

In a variation the reaction of phenols and beta-ketoesters and phosphorus pentoxide yields a chromone. This reaction is called Simonis chromone cyclization. The ketone in the ketoester is activated by P_2O_5 for reaction with the phenol hydroxyl group first, the ester group in it is then activated for electrophilic attack of the arene.

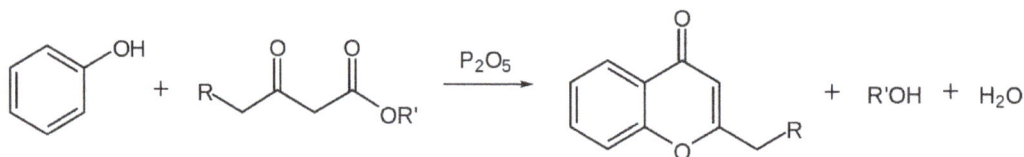

Heck Reaction

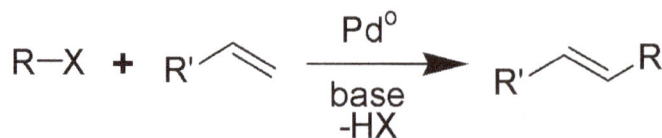

The Heck reaction

The Heck reaction (also called the Mizoroki-Heck reaction) is the chemical reaction of an unsaturated halide (or triflate) with an alkene in the presence of a base and a palladium catalyst (or palladium nanomaterial-based catalyst) to form a substituted alkene. It is named after Tsutomu Mizoroki and Richard F. Heck. Heck was awarded the 2010 Nobel Prize in Chemistry, which he shared with Ei-ichi Negishi and Akira Suzuki, for the discovery and development of this reaction. This reaction was the first example of a carbon-carbon bond-forming reaction that followed a Pd(0)/Pd(II) catalytic cycle, the same catalytic cycle that is seen in other Pd(0)-catalyzed cross-coupling reactions. The Heck reaction is of great importance, as it allows one to do substitution reactions on planar sp²-hybridized carbon atoms.

The reaction is performed in the presence of an organopalladium catalyst. The halide (Br, Cl) or triflate is an aryl, benzyl, or vinyl compound and the alkene contains at least one hydrogen and is often electron-deficient such as acrylate ester or an acrylonitrile.The catalyst can be tetrakis(triphenylphosphine) palladium(0), palladium chloride or palladium(II) acetate. The ligand is triphenylphosphine, PHOX or BINAP. The base is triethylamine, potassium carbonate or sodium acetate.

Several reviews have been published.

History

The original reaction by Tsutomu Mizoroki (1971) describes the coupling between iodobenzene and styrene to form stilbene in methanol at 120 °C (autoclave) with potassium acetate base and palladium chloride catalysis. This work was an extension of earlier work by Fujiwara (1967) on the Pd(II)-mediated coupling of arenes (Ar–H) and alkenes and earlier work by Heck (1969) on the coupling of arylmercuric halides (ArHgCl) with alkenes using a stoichiometric amount of a palladium(II) species.

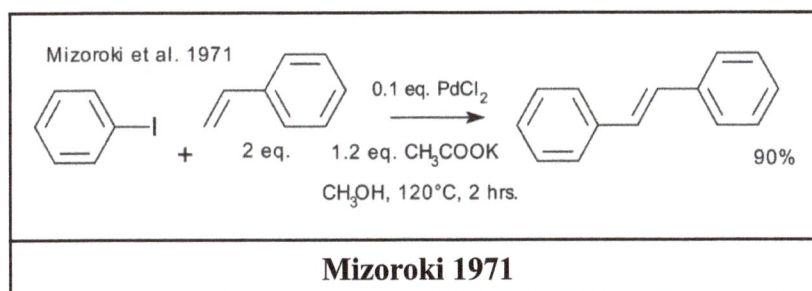

Mizoroki 1971

The 1972 Heck publication acknowledged the Mizoroki publication and detailed *independently discovered* work. The reaction conditions differ in catalyst used (palladium acetate) and catalyst loading (0.01 eq.), base used (a hindered amine) and lack of solvent.

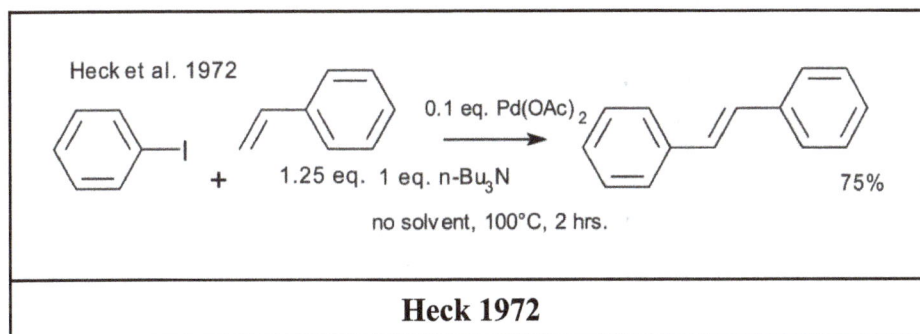

Heck 1972

In these reactions the active catalyst Pd(0) is formed by Pd coordination to the alkene.

In 1974 Heck introduced phosphine ligands into the equation.

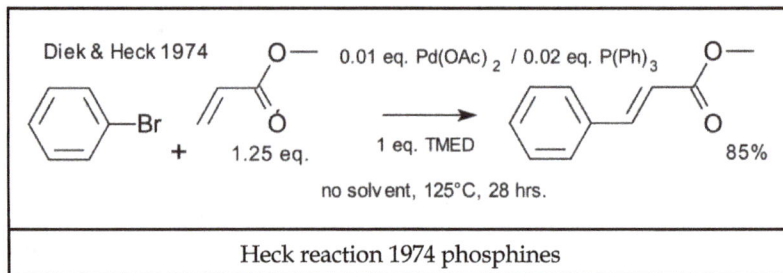

Heck reaction 1974 phosphines

Reaction Mechanism

The catalytic cycle for the Heck reaction involves a series of transformations around the palladium catalyst. The palladium(0) compound required in this cycle is generally prepared in situ from a palladium(II) precursor.

For instance, palladium(II) acetate is reduced by triphenylphosphine to bis(triphenylphosphine) palladium(0) (**1**) and triphenylphosphine is oxidized to triphenylphosphine oxide. Step **A** is an oxidative addition in which palladium inserts itself in the aryl to bromide bond. Palladium then forms a π complex with the alkene (**3**) and in step **B** the alkene inserts itself in the palladium - carbon bond in a syn addition step. Then follows a torsional strain relieving rotation to the trans isomer and step **C** is a beta-hydride elimination step with the formation of a new palladium - alkene π complex (**5**). This complex is destroyed in the next step. The palladium(0) compound is regenerated by reductive elimination of the palladium(II) compound by potassium carbonate in the final step, **D**. In the course of the reaction the carbonate is stoichiometrically consumed and palladium is truly a catalyst and used in catalytic amounts. A similar palladium cycle but with different scenes and actors is observed in the Wacker process.

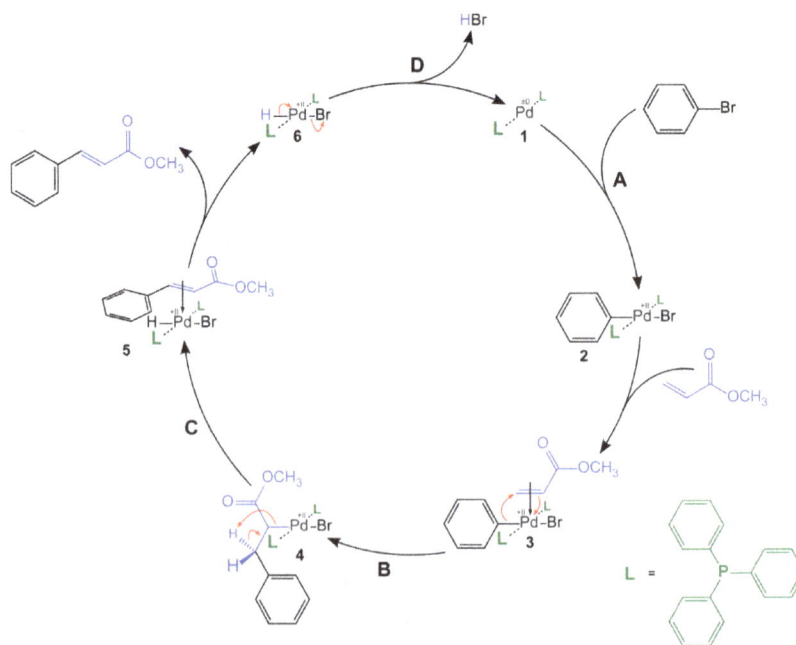

Heck Reaction Mechanism

This cycle is not limited to vinyl compounds, in the Sonogashira coupling one of the reactants is an alkyne and in the Suzuki coupling the alkene is replaced by an aryl boronic acid and in the Stille reaction by an aryl stannane. The cycle also extends to the other group 10 element nickel for example in the Negishi coupling between aryl halides and organozinc compounds. Platinum forms strong bonds with carbon and does not have a catalytic activity in this type of reaction.

Stereoselectivity

This coupling reaction is stereoselective with a propensity for trans coupling as the palladium halide group and the bulky organic residue move away from each other in the reaction sequence in a rotation step. The Heck reaction is applied industrially in the production of naproxen and the sunscreen component octyl methoxycinnamate. The naproxen synthesis includes a coupling between a brominated naphthalene compound with ethylene:

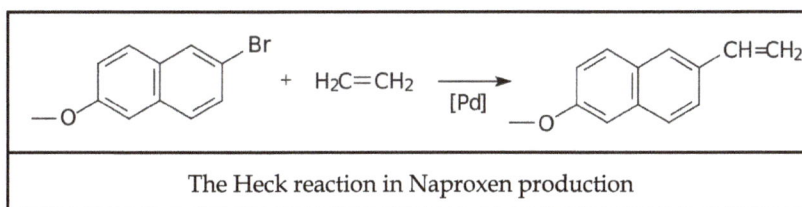

The Heck reaction in Naproxen production

Variations

Ionic Liquid Heck Reaction

In the presence of an ionic liquid a Heck reaction proceeds in absence of a phosphorus ligand. In one modification palladium acetate and the ionic liquid (bmim)PF_6 are immobilized inside the cavities of reversed-phase silica gel. In this way the reaction proceeds in water and the catalyst is re-usable.

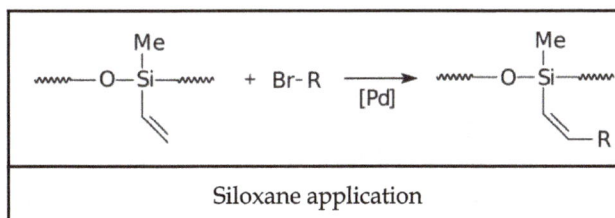

Siloxane application

Heck Oxyarylation

In the Heck oxyarylation modification the palladium substituent in the syn-addition intermediate is displaced by a hydroxyl group and the reaction product contains a dihydrofuran ring.

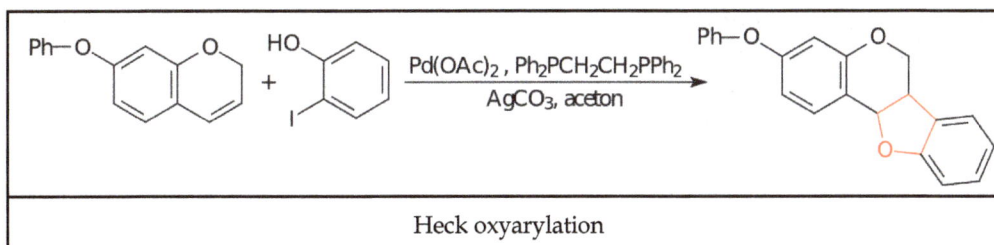

Heck oxyarylation

Amino-Heck Reaction

In the **amino-Heck reaction** a nitrogen to carbon bond is formed. In one example, an oxime with a strongly electron withdrawing group reacts intramolecularly with the terminal end of a diene to a pyridine compound. The catalyst is tetrakis(triphenylphosphine)palladium(0) and the base is triethylamine.

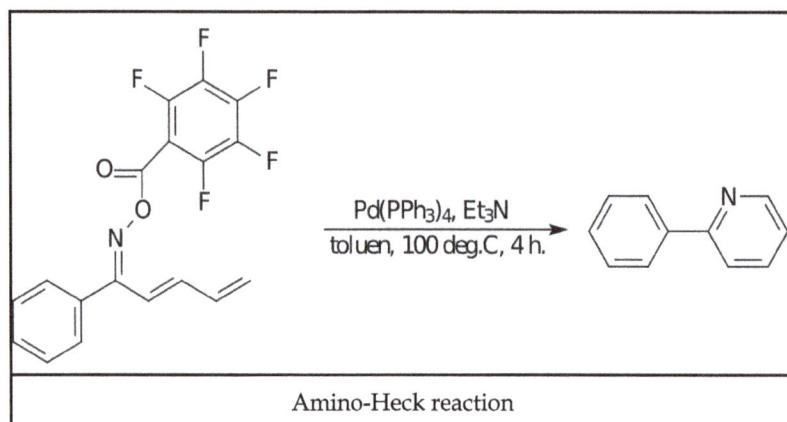

Amino-Heck reaction

References

- Kürti, László; Czakó, Barbara (2005). Strategic Applications of Named Reactions in Organic Synthesis. Burlington; San Diego; London: Elsevier Academic Press. p. 28. ISBN 978-0-12-369483-6.

- Jones, Jr., Maitland; Fleming, Steven A. (2010). Organic Chemistry (4th ed.). Canada: W. W. Norton & Company. p. 293. ISBN 978-0-393-93149-5.

- Zaugg, H. E.; Martin, W. B. (1965). "A-Amidoalkylations at Carbon". Org. React. 14: 88. doi:10.1002/0471264180. or014.02. ISBN 0471264180.

- Kappe, C. Oliver (2005) "The Biginelli Reaction", in: J. Zhu and H. Bienaymé (eds.): Multicomponent Reactions, Wiley-VCH, Weinheim, ISBN 978-3-527-30806-4.

- Kappe, C. O.; Stadler, A. (2004). "The Biginelli Dihydropyrimidine Synthesis". Organic Reactions. 63. doi:10.1002/0471264180.or063.01. ISBN 0471264180.

- Clayden, Jonathan; Greeves, Nick; Warren, Stuart; Wothers, Peter (2001). Organic Chemistry (1st ed.). Oxford University Press. ISBN 978-0-19-850346-0.Page 1067

- Boyd, Robert W.; Morrison, Robert (1992). Organic chemistry. Englewood Cliffs, N.J: Prentice Hall. pp. 241–242. ISBN 0-13-643669-2.

- Daru, János; Stirling, András (4 November 2011). "Mechanism of the Pechmann Reaction: A Theoretical Study". The Journal of Organic Chemistry. 76 (21): 8749–8755. doi:10.1021/jo201439u.

- Świzdor, Alina (2013). "Baeyer-Villiger Oxidation of Some C19 Steroids by Penicillium lanosocoeruleum". Molecules. 18 (11): 13812–13822. doi:10.3390/molecules181113812.

- Drahl, Carmen (May 17, 2010). "In Names, History And Legacy". Chemical and Engineering News. 88 (22): 31–33. doi:10.1021/cen-v088n020.p031. Retrieved June 4, 2011.

- Mc Cartney, Dennis; Guiry, Patrick J. (2011). "The asymmetric Heck and related reactions". Chemical Society Reviews. 40 (10): 5122–5150. doi:10.1039/C1CS15101K. PMID 21677934.

6

Bioproducts and its Types

Bioproducts are the materials and the chemicals that are obtained from renewable biological resources. The types of bioproducts that have been explained in the following section are bioenergy, biomass, biofuel, natural oil polyols, bio-based material etc. The topics discussed in the section are of great importance to broaden the existing knowledge on bioproducts.

Bioproducts

Bioproducts or bio-based products are materials, chemicals and energy derived from renewable biological resources.

Bioresources

Biological resources include agriculture, forestry, and biologically-derived waste, and there are many other renewable bioresource examples. One of the scientific terms used to denote renewable bioresources is lignocellulose. Lignocellulosic tissues are biologically-derived natural resources containing some of the main constituents of the natural world. 1) Holocellulose is the carbohydrate fraction of lignocellulose that includes cellulose, a common building block made of sugar (glucose) that is the most abundant biopolymer, as well as hemicellulose. 2) Lignin is the second most abundant biopolymer. Cellulose and lignin are two of the primary natural polymers used by plants to store energy as well as to give strength, as is the case in woody plant tissues. Other energy storage chemicals in plants include oils, waxes, fats, etc., and because these other plant compounds have distinct properties, they offer potential for a host of different bioproducts

Conventional bioproducts and emerging bioproducts are two broad categories used to categorize bioproducts. Examples of conventional bio-based products include building materials, pulp and paper, and forest products. Examples of emerging bioproducts or biobased products include biofuels, bioenergy, starch-based and cellulose-based ethanol, bio-based adhesives, biochemicals, bioplastics, etc. Emerging bioproducts are active subjects of research and development, and these efforts have developed significantly since the turn of the 20/21st century, in part driven by the price of traditional petroleum-based products, by the environmental impact of petroleum use, and by an interest in many countries to become independent from foreign sources of oil. Bioproducts derived from bioresources can replace much of the fuels, chemicals, plastics etc. that are currently derived from petroleum.

Bioproducts Engineering

Bioproducts engineering (also referred to as bioprocess engineering) refers to engineering of bio-products from renewable bioresources. This pertains to the design, development and implementation of processes, technologies for the sustainable manufacture of materials, chemicals and energy from renewable biological resources.

Also referred to as Bioprocess Engineering: Bioprocess Engineering is a specialization of Biotechnology, Chemical Engineering or Biological Engineering or of Agricultural Engineering. It deals with the design and development of equipment and processes for the manufacturing of products such as food, feed, pharmaceuticals, nutraceuticals, chemicals, and polymers and paper from biological materials. Bioprocees engineering is a conglomerate of mathematics, biology and industrial design,and consists of various spectrums like designing of Fermentors,study of fermentors (mode of operations etc.). It also deals with studying various biotechnological processes used in industries for large scale production of biological product for optimization of yield in the end product and the quality of end product. Bio process engineering may include the work of mechanical,electrical and industrial engineers to apply principles of their disciplines to processes based on using living cells or sub component of such cells

Also referred to as Bioresource Engineering: Bioresource engineering is related to the applications of biological engineering, chemical engineering and agricultural engineering usually based on biological and/or agricultural feedstocks. Bioresource engineering is more general and encompasses a wider range of technologies and various elements such as biomass, biological waste treatment, bioenergy, biotransformations and bioresource systems analysis, and technologies associated with Thermochemical conversion technologies: combustion, pyrolysis, gasification, catalysis, etc. Biochemical conversion technologies: aerobic methods, anaerobic digestion, microbial growth processes, enzymatic methods, composting Products: fibre, fuels, feedstocks, fertilisers, building materials, polymers and other industrial products Management: modelling, systems analysis, decisions, support systems. The impact of urbanization and increasing demand for land, food, and water presents engineers in a world with serious challenges. Little attention has been given to the interface between the biological world and traditional engineering in the past. It is the job of bioresource engineers to fill that gap. Agricultural and bioresource engineers develop efficient and environmentally-sensitive methods of producing food, fiber, timber, bio-based products and renewable energy sources for an ever-increasing world population.

Bioenergy

A Stirling engine capable of producing electricity from biomass combustion heat

Bioenergy is renewable energy made available from materials derived from biological sources. Biomass is any organic material which has stored sunlight in the form of chemical energy. As a fuel it may include wood, wood waste, straw, manure, sugarcane, and many other by-products from a variety of agricultural processes. By 2010, there was 35 GW (47,000,000 hp) of globally installed bioenergy capacity for electricity generation, of which 7 GW (9,400,000 hp) was in the United States.

In its most narrow sense it is a synonym to biofuel, which is fuel derived from biological sources. In its broader sense it includes biomass, the biological material used as a biofuel, as well as the social, economic, scientific and technical fields associated with using biological sources for energy. This is a common misconception, as bioenergy is the energy extracted from the biomass, as the biomass is the fuel and the bioenergy is the energy contained in the fuel

There is a slight tendency for the word *bioenergy* to be favoured in Europe compared with *biofuel* in America.

Solid Biomass

Simple use of biomass fuel (Combustion of wood for heat).

One of the advantages of biomass fuel is that it is often a by-product, residue or waste-product of other processes, such as farming, animal husbandry and forestry. In theory this means there is no competition between fuel and food production, although this is not always the case. Land use, existing biomass industries and relevant conversion technologies must be considered when evaluating suitability of developing biomass as feedstock for energy.

Biomass is the material derived from recently living organisms, which includes plants, animals and their byproducts. Manure, garden waste and crop residues are all sources of biomass. It is a renewable energy source based on the carbon cycle, unlike other natural resources such as petroleum, coal, and nuclear fuels. Another source includes Animal waste, which is a persistent and unavoidable pollutant produced primarily by the animals housed in industrial-sized farms.

There are also agricultural products specifically being grown for biofuel production. These include corn, and soybeans and to some extent willow and switchgrass on a pre-commercial research level, primarily in the United States; rapeseed, wheat, sugar beet, and willow (15,000 ha or 37,000 acres in Sweden) primarily in Europe; sugarcane in Brazil; palm oil and miscanthus in Southeast Asia; sorghum and cassava in China; and jatropha in India. Hemp has also been proven to work as a biofuel. Biodegradable outputs from industry, agriculture, forestry and households can be used for biofuel production, using e.g. anaerobic digestion to produce biogas, gasification to produce syngas or by direct combustion. Examples of biodegradable wastes include straw, timber, manure, rice husks, sewage, and food waste. The use of biomass fuels can therefore contribute to waste management as well as fuel security and help to prevent or slow down climate change, although alone they are not a comprehensive solution to these problems.

Biomass can be converted to other usable forms of energy like methane gas or transportation fuels like ethanol and biodiesel. Rotting garbage, and agricultural and human waste, all release methane gas—also called "landfill gas" or "biogas." Crops, such as corn and sugar cane, can be fermented to produce the transportation fuel, ethanol. Biodiesel, another transportation fuel, can be produced from left-over food products like vegetable oils and animal fats. Also, Biomass to liquids (BTLs) and cellulosic ethanol are still under research.

Sewage Biomass

A new bioenergy sewage treatment process aimed at developing countries is now on the horizon; the Omni Processor is a self-sustaining process which uses the sewerage solids as fuel to convert sewage waste water into drinking water and electrical energy.

Electricity Generation from Biomass

The biomass used for electricity production ranges by region. Forest byproducts, such as wood residues, are popular in the United States. Agricultural waste is common in Mauritius (sugar cane residue) and Southeast Asia (rice husks). Animal husbandry residues, such as poultry litter, is popular in the UK.

Electricity from Sugarcane Bagasse in Brazil

Sugarcane (*Saccharum officinarum*) plantation ready for harvest, Ituverava, São Paulo State. Brazil.

A sugar/ethanol plant located in Piracicaba, São Paulo State. This plant produces the electricity it needs from bagasse residuals from sugarcane left over by the milling process, and it sells the surplus electricity to the public grid.

Sucrose accounts for little more than 30% of the chemical energy stored in the mature plant; 35% is in the leaves and stem tips, which are left in the fields during harvest, and 35% are in the fibrous material (bagasse) left over from pressing.

The production process of sugar and ethanol in Brazil takes full advantage of the energy stored in sugarcane. Part of the bagasse is currently burned at the mill to provide heat for distillation and electricity to run the machinery. This allows ethanol plants to be energetically self-sufficient and even sell surplus electricity to utilities; current production is 600 MW (800,000 hp) for self-use and 100 MW (130,000 hp) for sale. This secondary activity is expected to boom now that utilities have been induced to pay "fair price "(about US$10/GJ or US$0.036/kWh) for 10 year contracts. This is approximately half of what the World Bank considers the reference price for investing in similar projects. The energy is especially valuable to utilities because it is produced mainly in the dry season when hydroelectric dams are running low. Estimates of potential power generation from bagasse range from 1,000 to 9,000 MW (1,300,000 to 12,100,000 hp), depending on technology. Higher estimates assume gasification of biomass, replacement of current low-pressure steam boilers and turbines by high-pressure ones, and use of harvest trash currently left behind in the fields. For comparison, Brazil's Angra I nuclear plant generates 657 MW (881,000 hp).

Presently, it is economically viable to extract about 288 MJ of electricity from the residues of one tonne of sugarcane, of which about 180 MJ are used in the plant itself. Thus a medium-size distillery processing 1,000,000 tonnes (980,000 long tons; 1,100,000 short tons) of sugarcane per year could sell about 5 MW (6,700 hp) of surplus electricity. At current prices, it would earn US$18 million from sugar and ethanol sales, and about US$1 million from surplus electricity sales. With advanced boiler and turbine technology, the electricity yield could be increased to 648 MJ per tonne of sugarcane, but current electricity prices do not justify the necessary investment. (According to one report, the World Bank would only finance investments in bagasse power generation if the price were at least US$19/GJ or US$0.068/kWh.)

Bagasse burning is environmentally friendly compared to other fuels like oil and coal. Its ash content is only 2.5% (against 30–50% of coal), and it contains very little sulfur. Since it burns at relatively low temperatures, it produces little nitrous oxides. Moreover, bagasse is being sold for use as a fuel (replacing heavy fuel oil) in various industries, including citrus juice concentrate, vegetable oil, ceramics, and tyre recycling. The state of São Paulo alone used 2,000,000 tonnes (1,970,000 long tons; 2,200,000 short tons), saving about US$35 million in fuel oil imports.

Researchers working with cellulosic ethanol are trying to make the extraction of ethanol from sugarcane bagasse and other plants viable on an industrial scale.

Electricity from Electrogenic Micro-organisms

Another form of bioenergy can be attained from microbial fuel cells, in which chemical energy stored in wastewater or soil is converted directly into electrical energy via the metabolic processes of electrogenic micro-organisms. The power generation capability of this technology has not been economical to date, however, and this technology has found more utility for chemical treatment processes and student education.

Environmental Impact

Some forms of forest bioenergy have recently come under fire from a number of environmental organizations, including Greenpeace and the Natural Resources Defense Council, for the harmful impacts they can have on forests and the climate. Greenpeace recently released a report entitled Fuelling a BioMess which outlines their concerns around forest bioenergy. Because any part of the tree can be burned, the harvesting of trees for energy production encourages Whole-Tree Harvesting, which removes more nutrients and soil cover than regular harvesting, and can be harmful to the long-term health of the forest. In some jurisdictions, forest biomass is increasingly consisting of elements essential to functioning forest ecosystems, including standing trees, naturally disturbed forests and remains of traditional logging operations that were previously left in the forest. Environmental groups also cite recent scientific research which has found that it can take many decades for the carbon released by burning biomass to be recaptured by regrowing trees, and even longer in low productivity areas; furthermore, logging operations may disturb forest soils and cause them to release stored carbon. In light of the pressing need to reduce greenhouse gas emissions in the short term in order to mitigate the effects of climate change, a number of environmental groups are opposing the large-scale use of forest biomass in energy production.

The New Scientist described a scenario in a September 2016 article which illustrated why the journal believed bioenergy can be bad: Suppose you cut down a 50 year oak tree in your garden and use the logs to heat your house instead of coal. Wood emits more carbon dioxide than coal per unit of heat gained and the roots left in the soil emit more carbon dioxide as they rot. If you plant another tree it will soak up that carbon dioxide in about 50 years. But if you had left the original tree in place it would have soaked up the carbon dioxide from the coal and more. It could take centuries before cutting down the tree would give any benefit. But the world needed to cut carbon dioxide over the next few decades if the global warming was to be kept below 3 degrees C. The journal also concluded that official claimed carbon reductions from renewables had been overstated. The European Union, for example, got more 64% of its renewable energy from biomass (mostly wood) but United Nations and EU rules did not count the carbon emissions from burning biomass.

Biomass

Sugarcane plantation in Brazil. Sugarcane bagasse is a type of biomass.

Biomass is organic matter derived from living, or recently living organisms. Biomass can be used as a source of energy and it most often refers to plants or plant-based materials that are not used for food or feed, and are specifically called lignocellulosic biomass. As an energy source, biomass can either be used directly via combustion to produce heat, or indirectly after converting it to various forms of biofuel. Conversion of biomass to biofuel can be achieved by different methods which are broadly classified into: *thermal*, *chemical*, and *biochemical* methods.

A cogeneration plant in Metz, France. The station uses waste wood biomass as an energy source, and provides electricity and heat for 30,000 dwellings.

Stump harvesting increases the recovery of biomass from a forest.

Biomass Sources

Historically, humans have harnessed biomass-derived energy since the time when people began burning wood to make fire. Even today, biomass is the only source of fuel for domestic use in many developing countries. Biomass is all biologically-produced matter based in carbon, hydrogen and oxygen. The estimated biomass production in the world is 104.9 petagrams ($104.9 * 10^{15}$ g – about 105 billion metric tons) of carbon per year, about half in the ocean and half on land.

Wood remains the largest biomass energy source today; examples include forest residues (such as dead trees, branches and tree stumps), yard clippings, wood chips and even municipal solid waste. Wood energy is derived by using lignocellulosic biomass (second-generation biofuels) as fuel. Harvested wood may be used directly as a fuel or collected from wood waste streams to be processed into pellet fuel or other forms of fuels. The largest source of energy from wood is pulping liquor or "black liquor," a waste product from processes of the pulp, paper and paperboard industry. In the second sense, biomass includes plant or animal matter that can be converted into fibers or other industrial chemicals, including biofuels. Industrial biomass can be grown from numerous types of plants, including miscanthus, switchgrass, hemp, corn, poplar, willow, sorghum, sugarcane, bamboo, and a variety of tree species, ranging from eucalyptus to oil palm (palm oil).

Eucalyptus in Brazil. Remains of the tree are reused for power generation.

Based on the source of biomass, biofuels are classified broadly into two major categories. First-generation biofuels are derived from sources such as sugarcane and corn starch. Sugars present in this biomass are fermented to produce bioethanol, an alcohol fuel which can be used directly in a fuel cell to produce electricity or serve as an additive to gasoline. However, utilizing food-based resources for fuel production only aggravates the food shortage problem. Second-generation biofuels, on the other hand, utilize non-food-based biomass sources such as agriculture and municipal waste. These biofuels mostly consist of lignocellulosic biomass, which is not edible and is a low-value waste for many industries. Despite being the favored alternative, economical production of second-generation biofuel is not yet achieved due to technological issues. These issues arise mainly due to chemical inertness and structural rigidity of lignocellulosic biomass.

Plant energy is produced by crops specifically grown for use as fuel that offer high biomass output per hectare with low input energy. Some examples of these plants are wheat, which typically yields 7.5–8 tonnes of grain per hectare, and straw, which typically yields 3.5–5 tonnes per hectare in the UK. The grain can be used for liquid transportation fuels while the straw can be burned to produce heat or electricity. Plant biomass can also be degraded from cellulose to glucose through a series of chemical treatments, and the resulting sugar can then be used as a first-generation biofuel.

The main contributors of waste energy are municipal solid waste, manufacturing waste, and landfill gas. Energy derived from biomass is projected to be the largest non-hydroelectric renewable resource of electricity in the US between 2000 and 2020.

Biomass can be converted to other usable forms of energy like methane gas or transportation fuels like ethanol and biodiesel. Rotting garbage, and agricultural and human waste, all release methane gas, also called landfill gas or biogas. Crops such as corn and sugarcane can be fermented to produce the transportation fuel ethanol. Biodiesel, another transportation fuel, can be produced from leftover food products like vegetable oils and animal fats. Several biodiesel companies simply collect used restaurant cooking oil and convert it into biodiesel. Also, biomass-to-liquids (called "BTLs") and cellulosic ethanol are still under research.

There is research involving algae, or algae-derived, biomass, as this non-food resource can be produced at rates five to ten times those of other types of land-based agriculture, such as corn and soy. Once harvested, it can be fermented to produce biofuels such as ethanol, butanol, and methane, as well as biodiesel and hydrogen. Efforts are being made to identify which species of algae are most suitable for energy production. Genetic engineering approaches could also be utilized to improve microalgae as a source of biofuel.

The biomass used for electricity generation varies by region. Forest by-products, such as wood residues, are common in the US. Agricultural waste is common in Mauritius (sugar cane residue) and Southeast Asia (rice husks). Animal husbandry residues, such as poultry litter, are common in the UK.

As of 2015, a new bioenergy sewage treatment process aimed at developing countries is under trial; the Omni Processor is a self-sustaining process which uses sewerage solids as fuel in a process to convert waste water into drinking water, with surplus electrical energy being generated for export.

Comparison of Total Plant Biomass Yields (Dry Basis)

World Resources

If the total annual primary production of biomass is just over 100 billion (1.0E+11) tonnes of Carbon /yr, and the energy reserve per metric tonne of biomass is between about 1.5E3 – 3E3 Kilowatt hours (5E6 – 10E6 BTU), or 24.8 TW average, then biomass could in principle provide 1.4 times the approximate annual 150E3 Terrawatt hours required for current world energy consumption. For reference, the total solar power on Earth is 174 kTW. The biomass equivalent to solar energy ratio is 143 ppm (parts per million), given current living system coverage on Earth. Best in class solar cell efficiency is (20–40)%. Additionally, Earth's internal radioactive energy production, largely the driver for volcanic activity, continental drift, etc., is in the same range of power, 20 TW. At some 50% carbon mass content in biomass, annual production, this corresponds to about 6% of atmospheric carbon content in CO_2 (for the current 400 ppm).

> (1.0E+11 tonnes biomass annually produced approximately 25 TW)

> Annual world biomass energy equivalent =16.7 – 33.4 TW.

> Annual world energy consumption =17.7. On average, biomass production is 1.4 times larger than world energy consumption.

Common Commodity Food Crops

- Agave: 1–21 tons/acre

- Alfalfa: 4–6 tons/acre

- Barley: grains – 1.6–2.8 tons/acre, straw – 0.9–2.5 tons/acre, total – 2.5–5.3 tons/acre

- Corn: grains – 3.2–4.9 tons/acre, stalks and stovers – 2.3–3.4 tons/acre, total – 5.5–8.3 tons/acre

- Jerusalem artichokes: tubers 1–8 tons/acre, tops 2–13 tons/acre, total 9–13 tons/acre

- Oats: grains – 1.4–5.4 tons/acre, straw – 1.9–3.2 tons/acre, total – 3.3–8.6 tons/acre

- Rye: grains – 2.1–2.4 tons/acre, straw – 2.4–3.4 tons/acre, total – 4.5–5.8 tons/acre

- Wheat: grains – 1.2–4.1 tons/acre, straw – 1.6–3.8 tons/acre, total – 2.8–7.9 tons/acre

Woody crops

- Oil palm: fronds 11 ton/acre, whole fruit bunches 1 ton/acre, trunks 30 ton/acre

Not yet in Commercial Planting

- Giant miscanthus: 5–15 tons/acre

- Sunn hemp: 4.5 tons/acre

- Switchgrass: 4–6 tons/acre

Genetically Modified Varieties

- Energy Sorghum

Thermal Conversion

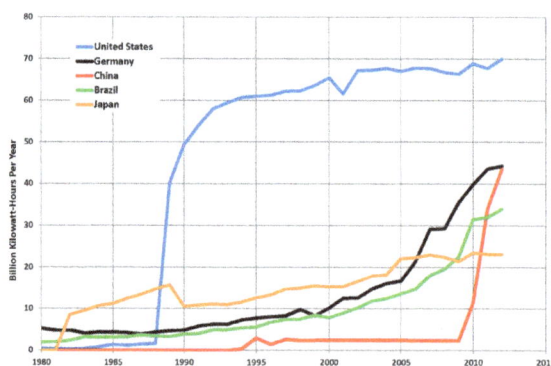

Trends in the top five countries generating electricity from biomass

Thermal conversion processes use heat as the dominant mechanism to convert biomass into another chemical form. Also known as thermal oil heating, it is a type of indirect heating in which a liquid phase heat transfer medium is heated and circulated to one or more heat energy users within a closed loop system. The basic alternatives of combustion (torrefaction, pyrolysis, and gasification) are separated principally by the extent to which the chemical reactions involved are allowed to proceed (mainly controlled by the availability of oxygen and conversion temperature).

Biomass briquettes are an example fuel for production of dendrothermal energy

Energy created by burning biomass (fuel wood) is particularly suited for countries where the fuel wood grows more rapidly, e.g. tropical countries. There are a number of other less common, more experimental or proprietary thermal processes that may offer benefits such as hydrothermal upgrading (HTU) and hydroprocessing. Some have been developed for use on high moisture content biomass, including aqueous slurries, and allow them to be converted into more convenient forms. Some of the applications of thermal conversion are combined heat and power (CHP) and co-firing. In a typical dedicated biomass power plant, efficiencies range from 20–27% (higher heating value basis). Biomass cofiring with coal, by contrast, typically occurs at efficiencies near those of the coal combustor (30–40%, higher heating value basis).

Chemical Conversion

A range of chemical processes may be used to convert biomass into other forms, such as to produce a fuel that is more conveniently used, transported or stored, or to exploit some property of the process itself. Many of these processes are based in large part on similar coal-based processes, such as Fischer-Tropsch synthesis, methanol production, olefins (ethylene and propylene), and similar chemical or fuel feedstocks. In most cases, the first step involves gasification, which step generally is the most expensive and involves the greatest technical risk. Biomass is more difficult to feed into a pressure vessel than coal or any liquid. Therefore, biomass gasification is frequently done at atmospheric pressure and causes combustion of biomass to produce a combustible gas consisting of carbon monoxide, hydrogen, and traces of methane. This gas mixture, called a producer gas, can provide fuel for various vital processes, such as internal combustion engines, as well as substitute for furnace oil in direct heat applications. Because any biomass material can undergo gasification, this process is far more attractive than ethanol or biomass production, where only particular biomass materials can be used to produce a fuel. In addition, biomass gasification is a desirable process due to the ease at which it can convert solid waste (such as wastes available on a farm) into producer gas, which is a very usable fuel.

Conversion of biomass to biofuel can also be achieved via selective conversion of individual components of biomass. For example, cellulose can be converted to intermediate platform chemical such a sorbitol, glucose, hydroxymethylfurfural etc. These chemical are then further reacted to produce hydrogen or hydrocarbon fuels.

Biomass also has the potential to be converted to multiple commodity chemicals. Halomethanes have successfully been by produced using a combination of A. fermentans and engineered S. cerevisiae. This method converts NaX salts and unprocessed biomass such as switchgrass, sugarcane, corn stover, or poplar into halomethanes. S-adenosylmethionine which is naturally occurring in S. cerevisiae allows a methyl group to be transferred. Production levels of 150 mg L-1H-1 iodomethane were achieved. At these levels roughly 173000L of capacity would need to be operated just to replace the United States' need for iodomethane. However, an advantage of this method is that it uses NaI rather than I2; NaI is significantly less hazardous than I2. This method may be applied to produce ethylene in the future.

Other chemical processes such as converting straight and waste vegetable oils into biodiesel is transesterification.

Biochemical Conversion

As biomass is a natural material, many highly efficient biochemical processes have developed in nature to break down the molecules of which biomass is composed, and many of these biochemical conversion processes can be harnessed.

Biochemical conversion makes use of the enzymes of bacteria and other microorganisms to break down biomass into gaseous or liquid fuels, such a biogas or bioethanol. In most cases, microorganisms are used to perform the conversion process: anaerobic digestion, fermentation, and composting.

Electrochemical Conversion

In addition to combustion, biomass/biofuels can be directly converted to electrical energy via electrochemical oxidation of the material. This can be performed directly in a direct carbon fuel cell, direct ethanol fuel cell or a microbial fuel cell. The fuel can also be consumed indirectly via a fuel cell system containing a reformer which converts the biomass into a mixture of CO and H2 before it is consumed in the fuel cell.

In the United States

The biomass power generating industry in the United States consists of approximately 11,000 MW of summer operating capacity actively supplying power to the grid, and produces about 1.4 percent of the U.S. electricity supply.

Public Service of New Hampshire (later merged with other companies into Eversource) in 2006 replaced a 50 MW coal boiler with a new 50 MW biomass boiler at its Schiller Station facility in Portsmouth, NH. The boiler's biomass fuel is from sources in NH, Massachusetts and Maine.

Currently, the New Hope Power Partnership is the largest biomass power plant in the U.S. The 140 MW facility uses sugarcane fiber (bagasse) and recycled urban wood as fuel to generate enough power for its large milling and refining operations as well as to supply electricity for nearly 60,000 homes.

Second-Generation Biofuels

Second-generation biofuels were not (in 2010) produced commercially, but a considerable number of research activities were taking place mainly in North America, Europe and also in some emerging countries. These tend to use feedstock produced by rapidly reproducing enzymes or bacteria from various sources including excrement grown in Cell cultures or hydroponics There is huge potential for second generation biofuels but non-edible feedstock resources are highly under-utilized.

Environmental Impact

Using biomass as a fuel produces air pollution in the form of carbon monoxide, carbon dioxide, NOx (nitrogen oxides), VOCs (volatile organic compounds), particulates and other pollutants at levels above those from traditional fuel sources such as coal or natural gas in some cases (such as with indoor heating and cooking). Utilization of wood biomass as a fuel can also produce fewer particulate and other pollutants than open burning as seen in wildfires or direct heat applications. Black carbon – a pollutant created by combustion of fossil fuels, biofuels, and biomass – is possibly the second largest contributor to global warming. In 2009 a Swedish study of the giant brown haze that periodically covers large areas in South Asia determined that it had been principally produced by open burning of biomass, and to a lesser extent by fossil-fuel burning. Researchers measured a significant concentration of ^{14}C (Carbon-14), which is associated with recent plant life rather than with fossil fuels.

Biomass power plant size is often driven by biomass availability in close proximity as transport costs of the (bulky) fuel play a key factor in the plant's economics. It has to be noted, however, that rail and especially shipping on waterways can reduce transport costs significantly, which has led to a global biomass market. To make small plants of 1 MW_{el} economically profitable those power plants need to be equipped with technology that is able to convert biomass to useful electricity with high efficiency such as ORC technology, a cycle similar to the water steam power process just with an organic working medium. Such small power plants can be found in Europe.

On combustion, the carbon from biomass is released into the atmosphere as carbon dioxide (CO_2). The amount of carbon stored in dry wood is approximately 50% by weight. However, according to the Food and Agriculture Organization of the United Nations, plant matter used as a fuel can be replaced by planting for new growth. When the biomass is from forests, the time to recapture the carbon stored is generally longer, and the carbon storage capacity of the forest may be reduced overall if destructive forestry techniques are employed.

Industry professionals claim that a range of issues can affect a plant's ability to comply with emissions standards. Some of these challenges, unique to biomass plants, include inconsistent fuel supplies and age. The type and amount of the fuel supply are completely reliant factors; the fuel can be in the form of building debris or agricultural waste (such as removal of invasive species or orchard trimmings). Furthermore, many of the biomass plants are old, use outdated technology and were not built to comply with today's stringent standards. In fact, many are based on technologies developed during the term of U.S. President Jimmy Carter, who created the United States Department of Energy in 1977.

The U.S. Energy Information Administration projected that by 2017, biomass is expected to be about twice as expensive as natural gas, slightly more expensive than nuclear power, and much less expensive than solar panels. In another EIA study released, concerning the government's plan to implement a 25% renewable energy standard by 2025, the agency assumed that 598 million tons of biomass would be available, accounting for 12% of the renewable energy in the plan.

The adoption of biomass-based energy plants has been a slow but steady process. Between the years of 2002 and 2012 the production of these plants has increased 14%. In the United States, alternative electricity-production sources on the whole generate about 13% of power; of this fraction, biomass contributes approximately 11% of the alternative production. According to a study conducted in early 2012, of the 107 operating biomass plants in the United States, 85 have been cited by federal or state regulators for the violation of clean air or water standards laws over the past 5 years. This data also includes minor infractions.

Despite harvesting, biomass crops may sequester carbon. For example, soil organic carbon has been observed to be greater in switchgrass stands than in cultivated cropland soil, especially at depths below 12 inches. The grass sequesters the carbon in its increased root biomass. Typically, perennial crops sequester much more carbon than annual crops due to much greater non-harvested living biomass, both living and dead, built up over years, and much less soil disruption in cultivation.

The proposal that biomass is carbon-neutral put forward in the early 1990s has been superseded by more recent science that recognizes that mature, intact forests sequester carbon more effectively than cut-over areas. When a tree's carbon is released into the atmosphere in a single pulse, it contributes to climate change much more than woodland timber rotting slowly over decades. Current studies indicate that "even after 50 years the forest has not recovered to its initial carbon storage" and "the optimal strategy is likely to be protection of the standing forest".

The pros and cons of biomass usage regarding carbon emissions may be quantified with the ILUC factor. There is controversy surrounding the usage of the ILUC factor.

Forest-based biomass has recently come under fire from a number of environmental organizations, including Greenpeace and the Natural Resources Defense Council, for the harmful impacts it can have on forests and the climate. Greenpeace recently released a report entitled "Fuelling a BioMess" which outlines their concerns around forest-based biomass. Because any part of the tree can be burned, the harvesting of trees for energy production encourages Whole-Tree Harvesting, which removes more nutrients and soil cover than regular harvesting, and can be harmful to the long-term health of the forest. In some jurisdictions, forest biomass removal is increasingly involving elements essential to functioning forest ecosystems, including standing trees, naturally disturbed forests and remains of traditional logging operations that were previously left in the forest. Environmental groups also cite recent scientific research which has found that it can take many decades for the carbon released by burning biomass to be recaptured by regrowing trees, and even longer in low productivity areas; furthermore, logging operations may disturb forest soils and cause them to release stored carbon. In light of the pressing need to reduce greenhouse gas emissions in the short term in order to mitigate the effects of climate change, a number of environmental groups are opposing the large-scale use of forest biomass in energy production.

Supply Chain Issues

With the seasonality of biomass supply and a great variability in sources, supply chains play a key role in cost-effective delivery of bioenergy. There are several potential challenges unique to bioenergy supply chains:

Technical issues

- Inefficiencies of the conversion processes

- Storage methods for seasonal availability

- Complex multi-component constituents incompatible with maximizing efficiency of single purpose use

- High water content of many biomass feedstock

- Conflicting decisions (technologies, locations, and routes)

- Complex location analysis (source points, inventory facilities, and production plants)

Logistic issues

- Seasonal availability leading to storage challenges and/or seasonally idle facilities

- Low bulk-density and/or high water content making transportation of biomass less economical

- Finite productivity per area and/or time incompatible with conventional approach to economy of scale focusing on maximizing facility size

Financial issues

- The limits for the traditional approach to economy of scale which focuses on maximizing single facility size

- Unavailability and complexity of life cycle costing data

- Lack of required transport infrastructure

- Limited flexibility or inflexibility to energy demand

- Risks associated with new technologies (insurability, performance, rate of return)

- Extended market volatilities (conflicts with alternative markets for biomass)

- Difficult or impossible to use financial hedging methods to control cost

Social issues

- Lack of participatory decision making

- Lack of public/community awareness

- Local supply chain impacts vs. global benefits

- Health and safety risks

- Extra pressure on transport sector

- Decreasing the esthetics of rural areas

Policy and regulatory issues

- Impact of fossil fuel tax on biomass transport

- Lack of incentives to create competition among bioenergy producers

- Focus on technology options and less attention to selection of biomass materials

- Lack of support for sustainable supply chain solutions

Institutional and organizational issues

- Varied ownership arrangements and priorities among supply chain parties

- Lack of supply chain standards

- Impact of organizational norms and rules on decision making and supply chain coordination

- Immaturity of change management practices in biomass supply chains

Biofuel

A bus fueled by biodiesel

A biofuel is a fuel that is produced through contemporary biological processes, such as agriculture and anaerobic digestion, rather than a fuel produced by geological processes such as those involved in the formation of fossil fuels, such as coal and petroleum, from prehistoric biological matter. Biofuels can be derived directly from plants, or indirectly from agricultural, commercial, domestic, and/or industrial wastes. Renewable biofuels generally involve contemporary carbon fixation, such as those that occur in plants or microalgae through the process of photosynthesis. Other renewable biofuels are made through the use or conversion of biomass (referring to recently living organisms, most often referring to plants or plant-derived materials). This biomass can be converted to convenient energy-containing substances in three different ways: thermal conversion, chemical conversion, and biochemical conversion. This biomass conversion can result in fuel in solid, liquid, or gas form. This new biomass can also be used directly for biofuels.

Information on pump regarding ethanol fuel blend up to 10%, California

Bioethanol is an alcohol made by fermentation, mostly from carbohydrates produced in sugar or starch crops such as corn, sugarcane, or sweet sorghum. Cellulosic biomass, derived from non-food sources, such as trees and grasses, is also being developed as a feedstock for ethanol production. Ethanol can be used as a fuel for vehicles in its pure form, but it is usually used as a gasoline additive to increase octane and improve vehicle emissions. Bioethanol is widely used in the USA and in Brazil. Current plant design does not provide for converting the lignin portion of plant raw materials to fuel components by fermentation.

Biodiesel can be used as a fuel for vehicles in its pure form, but it is usually used as a diesel additive to reduce levels of particulates, carbon monoxide, and hydrocarbons from diesel-powered vehicles. Biodiesel is produced from oils or fats using transesterification and is the most common biofuel in Europe.

In 2010, worldwide biofuel production reached 105 billion liters (28 billion gallons US), up 17% from 2009, and biofuels provided 2.7% of the world's fuels for road transport. Global ethanol fuel production reached 86 billion liters (23 billion gallons US) in 2010, with the United States and Brazil as the world's top producers, accounting together for about 90% of global production. The world's largest biodiesel producer is the European Union, accounting for 53% of all biodiesel production in 2010. As of 2011, mandates for blending biofuels exist in 31 countries at the national level and in 29 states or provinces. The International Energy Agency has a goal for biofuels to meet more than a quarter of world demand for transportation fuels by 2050 to reduce dependence on petroleum and coal. The production of biofuels also led into a flourishing automotive industry, where by 2010, 79% of all cars produced in Brazil were made with a hybrid fuel system of bioethanol and gasoline.

There are various social, economic, environmental and technical issues relating to biofuels production and use, which have been debated in the popular media and scientific journals. These include: the effect of moderating oil prices, the "food vs fuel" debate, poverty reduction potential, carbon emissions levels, sustainable biofuel production, deforestation and soil erosion, loss of biodiversity, impact on water resources, rural social exclusion and injustice, shantytown migration, rural unskilled unemployment, and nitrous oxide (NO_2) emissions.

Liquid Fuels for Transportation

Most transportation fuels are liquids, because vehicles usually require high energy density. This occurs naturally in liquids and solids. High energy density can also be provided by an internal combustion engine. These engines require clean-burning fuels. The fuels that are easiest to burn cleanly are typically liquids and gases. Thus, liquids meet the requirements of being both energy-dense and clean-burning. In addition, liquids (and gases) can be pumped, which means handling is easily mechanized, and thus less laborious.

First-generation Biofuels

"First-generation" or conventional biofuels are made from sugar, starch, or vegetable oil.

Ethanol

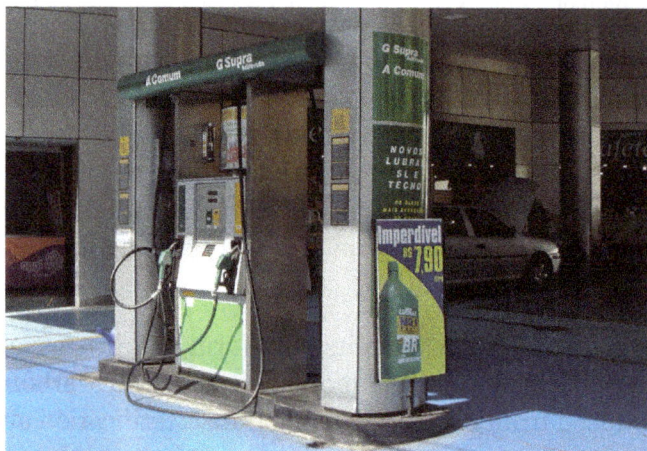

Neat ethanol on the left (A), gasoline on the right (G) at a filling station in Brazil

Biologically produced alcohols, most commonly ethanol, and less commonly propanol and butanol, are produced by the action of microorganisms and enzymes through the fermentation of sugars or starches (easiest), or cellulose (which is more difficult). Biobutanol (also called biogasoline) is often claimed to provide a direct replacement for gasoline, because it can be used directly in a gasoline engine.

Ethanol fuel is the most common biofuel worldwide, particularly in Brazil. Alcohol fuels are produced by fermentation of sugars derived from wheat, corn, sugar beets, sugar cane, molasses and any sugar or starch from which alcoholic beverages such as whiskey, can be made (such as potato and fruit waste, etc.). The ethanol production methods used are enzyme digestion (to release sugars from stored starches), fermentation of the sugars, distillation and drying. The distillation process requires significant energy input for heat (sometimes unsustainable natural gas fossil fuel, but cellulosic biomass such as bagasse, the waste left after sugar cane is pressed to extract its juice, is the most common fuel in Brazil, while pellets, wood chips and also waste heat are more common in Europe) Waste steam fuels ethanol factory - where waste heat from the factories also is used in the district heating grid.

U.S. President George W. Bush looks at sugar cane, a source of biofuel, with Brazilian President Luiz Inácio Lula da Silva during a tour on biofuel technology at Petrobras in São Paulo, Brazil, 9 March 2007.

Ethanol can be used in petrol engines as a replacement for gasoline; it can be mixed with gasoline to any percentage. Most existing car petrol engines can run on blends of up to 15% bioethanol with petroleum/gasoline. Ethanol has a smaller energy density than that of gasoline; this means it takes more fuel (volume and mass) to produce the same amount of work. An advantage of ethanol (CH_3CH_2OH) is that it has a higher octane rating than ethanol-free gasoline available at roadside gas stations, which allows an increase of an engine's compression ratio for increased thermal efficiency. In high-altitude (thin air) locations, some states mandate a mix of gasoline and ethanol as a winter oxidizer to reduce atmospheric pollution emissions.

Ethanol is also used to fuel bioethanol fireplaces. As they do not require a chimney and are "flue-less", bioethanol fires are extremely useful for newly built homes and apartments without a flue. The downsides to these fireplaces is that their heat output is slightly less than electric heat or gas fires, and precautions must be taken to avoid carbon monoxide poisoning.

Corn-to-ethanol and other food stocks has led to the development of cellulosic ethanol. According to a joint research agenda conducted through the US Department of Energy, the fossil energy ratios (FER) for cellulosic ethanol, corn ethanol, and gasoline are 10.3, 1.36, and 0.81, respectively.

Ethanol has roughly one-third lower energy content per unit of volume compared to gasoline. This is partly counteracted by the better efficiency when using ethanol (in a long-term test of more than 2.1 million km, the BEST project found FFV vehicles to be 1-26 % more energy efficient than petrol cars, but the volumetric consumption increases by approximately 30%, so more fuel stops are required.

With current subsidies, ethanol fuel is slightly cheaper per distance traveled in the United States.

Biodiesel

Biodiesel is the most common biofuel in Europe. It is produced from oils or fats using transesterification and is a liquid similar in composition to fossil/mineral diesel. Chemically, it consists mostly of fatty acid methyl (or ethyl) esters (FAMEs). Feedstocks for biodiesel include animal fats, vegetable oils, soy, rapeseed, jatropha, mahua, mustard, flax, sunflower, palm oil, hemp, field pennycress, *Pongamia pinnata* and algae. Pure biodiesel (B100) currently reduces emissions with up to 60% compared to diesel Second generation B100.

Biodiesel can be used in any diesel engine when mixed with mineral diesel. In some countries, manufacturers cover their diesel engines under warranty for B100 use, although Volkswagen of Germany, for example, asks drivers to check by telephone with the VW environmental services department before switching to B100. B100 may become more viscous at lower temperatures, depending on the feedstock used. In most cases, biodiesel is compatible with diesel engines from 1994 onwards, which use 'Viton' (by DuPont) synthetic rubber in their mechanical fuel injection systems. Note however, that no vehicles are certified for using neat biodiesel before 2014, as there was no emission control protocol available for biodiesel before this date.

Electronically controlled 'common rail' and 'unit injector' type systems from the late 1990s onwards may only use biodiesel blended with conventional diesel fuel. These engines have finely metered and atomized multiple-stage injection systems that are very sensitive to the viscosity of the fuel. Many current-generation diesel engines are made so that they can run on B100 without altering the engine itself, although this depends on the fuel rail design. Since biodiesel is an effective solvent and cleans residues deposited by mineral diesel, engine filters may need to be replaced more often, as the biofuel dissolves old deposits in the fuel tank and pipes. It also effectively cleans the engine combustion chamber of carbon deposits, helping to maintain efficiency. In many European countries, a 5% biodiesel blend is widely used and is available at thousands of gas stations. Biodiesel is also an oxygenated fuel, meaning it contains a reduced amount of carbon and higher hydrogen and oxygen content than fossil diesel. This improves the combustion of biodiesel and reduces the particulate emissions from unburnt carbon. However, using neat biodiesel may increase NOx-emissions

Biodiesel is also safe to handle and transport because it is non-toxic and biodegradable, and has a high flash point of about 300 °F (148 °C) compared to petroleum diesel fuel, which has a flash point of 125 °F (52 °C).

In the USA, more than 80% of commercial trucks and city buses run on diesel. The emerging US biodiesel market is estimated to have grown 200% from 2004 to 2005. "By the end of 2006 biodiesel production was estimated to increase fourfold [from 2004] to more than" 1 billion US gallons (3,800,000 m³).

In France, biodiesel is incorporated at a rate of 8% in the fuel used by all French diesel vehicles.

Avril Group produces under the brand Diester, a fifth of 11 million tons of biodiesel consumed annually by the European Union. It is the leading European producer of biodiesel.

Other Bioalcohols

Methanol is currently produced from natural gas, a non-renewable fossil fuel. In the future it is hoped to be produced from biomass as biomethanol. This is technically feasible, but the production is currently being postponed for concerns of Jacob S. Gibbs and Brinsley Coleberd that the economic viability is still pending. The methanol economy is an alternative to the hydrogen economy, compared to today's hydrogen production from natural gas.

Butanol (C_4H_9OH) is formed by ABE fermentation (acetone, butanol, ethanol) and experimental modifications of the process show potentially high net energy gains with butanol as the only liquid product. Butanol will produce more energy and allegedly can be burned "straight" in existing gasoline engines (without modification to the engine or car), and is less corrosive and less water-soluble than ethanol, and could be distributed via existing infrastructures. DuPont and BP are working together to help develop butanol. *E. coli* strains have also been successfully engineered to produce butanol by modifying their amino acid metabolism.

Green Diesel

Green diesel is produced through hydrocracking biological oil feedstocks, such as vegetable oils and animal fats. Hydrocracking is a refinery method that uses elevated temperatures and pressure in the presence of a catalyst to break down larger molecules, such as those found in vegetable oils, into shorter hydrocarbon chains used in diesel engines. It may also be called renewable diesel, hydrotreated vegetable oil or hydrogen-derived renewable diesel. Green diesel has the same chemical properties as petroleum-based diesel. It does not require new engines, pipelines or infrastructure to distribute and use, but has not been produced at a cost that is competitive with petroleum. Gasoline versions are also being developed. Green diesel is being developed in Louisiana and Singapore by ConocoPhillips, Neste Oil, Valero, Dynamic Fuels, and Honeywell UOP as well as Preem in Gothenburg, Sweden, creating what is known as Evolution Diesel.

Biofuel Gasoline

In 2013 UK researchers developed a genetically modified strain of Escherichia coli (E.Coli), which could transform glucose into biofuel gasoline that does not need to be blended. Later in 2013 UCLA researchers engineered a new metabolic pathway to bypass glycolysis and increase the rate of conversion of sugars into biofuel, while KAIST researchers developed a strain capable of producing short-chain alkanes, free fatty acids, fatty esters and fatty alcohols through the fatty acyl (acyl carrier protein (ACP)) to fatty acid to fatty acyl-CoA pathway *in vivo*. It is believed that in the future it will be possible to "tweak" the genes to make gasoline from straw or animal manure.

Vegetable Oil

Straight unmodified edible vegetable oil is generally not used as fuel, but lower-quality oil can and has been used for this purpose. Used vegetable oil is increasingly being processed into biodiesel, or (more rarely) cleaned of water and particulates and used as a fuel.

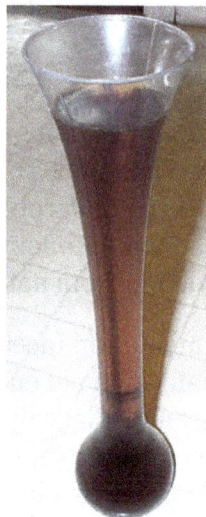

Filtered waste vegetable oil

Walmart's truck fleet logs millions of miles each year, and the company planned to double the fleet's efficiency between 2005 and 2015. This truck is one of 15 based at Walmart's Buckeye, Arizona distribution center that was converted to run on a biofuel made from reclaimed cooking grease produced during food preparation at Walmart stores.

As with 100% biodiesel (B100), to ensure the fuel injectors atomize the vegetable oil in the correct pattern for efficient combustion, vegetable oil fuel must be heated to reduce its viscosity to that of diesel, either by electric coils or heat exchangers. This is easier in warm or temperate climates. MAN B&W Diesel, Wärtsilä, and Deutz AG, as well as a number of smaller companies, such as Elsbett, offer engines that are compatible with straight vegetable oil, without the need for after-market modifications.

Vegetable oil can also be used in many older diesel engines that do not use common rail or unit injection electronic diesel injection systems. Due to the design of the combustion chambers in indirect injection engines, these are the best engines for use with vegetable oil. This system allows the relatively larger oil molecules more time to burn. Some older engines, especially Mercedes, are driven experimentally by enthusiasts without any conversion, a handful of drivers have experienced limited success with earlier pre-"Pumpe Duse" VW TDI engines and other similar engines with direct injection. Several companies, such as Elsbett or Wolf, have developed professional conversion kits and successfully installed hundreds of them over the last decades.

Oils and fats can be hydrogenated to give a diesel substitute. The resulting product is a straight-chain hydrocarbon with a high cetane number, low in aromatics and sulfur and does not contain oxygen. Hydrogenated oils can be blended with diesel in all proportions. They have several advantages over biodiesel, including good performance at low temperatures, no storage stability problems and no susceptibility to microbial attack.

Bioethers

Bioethers (also referred to as fuel ethers or oxygenated fuels) are cost-effective compounds that act as octane rating enhancers."Bioethers are produced by the reaction of reactive iso-olefins, such as iso-butylene, with bioethanol." Bioethers are created by wheat or sugar beet. They also enhance engine performance, whilst significantly reducing engine wear and toxic exhaust emissions. Though bioethers are likely to replace petroethers in the UK, it is highly unlikely they will become a fuel in and of itself due to the low energy density. Greatly reducing the amount of ground-level ozone emissions, they contribute to air quality.

When it comes to transportation fuel there are six ether additives: dimethyl ether (DME), diethyl ether (DEE), methyl teritiary-butyl ether (MTBE), ethyl *ter*-butyl ether (ETBE), t*er*-amyl methyl ether (TAME), and *ter*-amyl ethyl ether (TAEE)

The European Fuel Oxygenates Association (EFOA) credits methyl Ttertiary-butyl ether (MTBE) and ethyl ter-butyl ether (ETBE) as the most commonly used ethers in fuel to replace lead. Ethers were introduced in Europe in the 1970s to replace the highly toxic compound. Although Europeans still use bio-ether additives, the US no longer has an oxygenate requirement therefore bio-ethers are no longer used as the main fuel additive.

Biogas

Pipes carrying biogas

Biogas is methane produced by the process of anaerobic digestion of organic material by anaerobes. It can be produced either from biodegradable waste materials or by the use of energy crops fed into anaerobic digesters to supplement gas yields. The solid byproduct, digestate, can be used as a biofuel or a fertilizer.

Biogas can be recovered from mechanical biological treatment waste processing systems. Landfill gas, a less clean form of biogas, is produced in landfills through naturally occurring anaerobic digestion. If it escapes into the atmosphere, it is a potential greenhouse gas.

Farmers can produce biogas from manure from their cattle by using anaerobic digesters.

Syngas

Syngas, a mixture of carbon monoxide, hydrogen and other hydrocarbons, is produced by partial combustion of biomass, that is, combustion with an amount of oxygen that is not sufficient to convert the biomass completely to carbon dioxide and water. Before partial combustion, the biomass is dried, and sometimes pyrolysed. The resulting gas mixture, syngas, is more efficient than direct combustion of the original biofuel; more of the energy contained in the fuel is extracted.

Syngas may be burned directly in internal combustion engines, turbines or high-temperature fuel cells. The wood gas generator, a wood-fueled gasification reactor, can be connected to an internal combustion engine.

Syngas can be used to produce methanol, DME and hydrogen, or converted via the Fischer-Tropsch process to produce a diesel substitute, or a mixture of alcohols that can be blended into gasoline. Gasification normally relies on temperatures greater than 700 °C.

Lower-temperature gasification is desirable when co-producing biochar, but results in syngas polluted with tar.

Solid Biofuels

Examples include wood, sawdust, grass trimmings, domestic refuse, charcoal, agricultural waste, nonfood energy crops, and dried manure.

When raw biomass is already in a suitable form (such as firewood), it can burn directly in a stove or furnace to provide heat or raise steam. When raw biomass is in an inconvenient form (such as sawdust, wood chips, grass, urban waste wood, agricultural residues), the typical process is to densify the biomass. This process includes grinding the raw biomass to an appropriate particulate size (known as hogfuel), which, depending on the densification type, can be from 1 to 3 cm (0.4 to 1.2 in), which is then concentrated into a fuel product. The current processes produce wood pellets, cubes, or pucks. The pellet process is most common in Europe, and is typically a pure wood product. The other types of densification are larger in size compared to a pellet, and are compatible with a broad range of input feedstocks. The resulting densified fuel is easier to transport and feed into thermal generation systems, such as boilers.

Industry has used sawdust, bark and chips for fuel for decades, primary in the pulp and paper industry, and also bagasse (spent sugar cane) fueled boilers in the sugar cane industry. Boilers in the range of 500,000 lb/hr of steam, and larger, are in routine operation, using grate, spreader stoker, suspension burning and fluid bed combustion. Utilities generate power, typically in the range of

5 to 50 MW, using locally available fuel. Other industries have also installed wood waste fueled boilers and dryers in areas with low cost fuel.

One of the advantages of biomass fuel is that it is often a byproduct, residue or waste-product of other processes, such as farming, animal husbandry and forestry. In theory, this means fuel and food production do not compete for resources, although this is not always the case.

A problem with the combustion of raw biomass is that it emits considerable amounts of pollutants, such as particulates and polycyclic aromatic hydrocarbons. Even modern pellet boilers generate much more pollutants than oil or natural gas boilers. Pellets made from agricultural residues are usually worse than wood pellets, producing much larger emissions of dioxins and chlorophenols.

In spite of the above noted study, numerous studies have shown biomass fuels have significantly less impact on the environment than fossil based fuels. Of note is the US Department of Energy Laboratory, operated by Midwest Research Institute Biomass Power and Conventional Fossil Systems with and without CO2 Sequestration – Comparing the Energy Balance, Greenhouse Gas Emissions and Economics Study. Power generation emits significant amounts of greenhouse gases (GHGs), mainly carbon dioxide (CO_2). Sequestering CO_2 from the power plant flue gas can significantly reduce the GHGs from the power plant itself, but this is not the total picture. CO_2 capture and sequestration consumes additional energy, thus lowering the plant's fuel-to-electricity efficiency. To compensate for this, more fossil fuel must be procured and consumed to make up for lost capacity.

Taking this into consideration, the global warming potential (GWP), which is a combination of CO_2, methane (CH_4), and nitrous oxide (N_2O) emissions, and energy balance of the system need to be examined using a life cycle assessment. This takes into account the upstream processes which remain constant after CO_2 sequestration, as well as the steps required for additional power generation. Firing biomass instead of coal led to a 148% reduction in GWP.

A derivative of solid biofuel is biochar, which is produced by biomass pyrolysis. Biochar made from agricultural waste can substitute for wood charcoal. As wood stock becomes scarce, this alternative is gaining ground. In eastern Democratic Republic of Congo, for example, biomass briquettes are being marketed as an alternative to charcoal to protect Virunga National Park from deforestation associated with charcoal production.

Second-generation (Advanced) Biofuels

Second generation biofuels, also known as advanced biofuels, are fuels that can be manufactured from various types of biomass. Biomass is a wide-ranging term meaning any source of organic carbon that is renewed rapidly as part of the carbon cycle. Biomass is derived from plant materials but can also include animal materials.

First generation biofuels are made from the sugars and vegetable oils found in arable crops, which can be easily extracted using conventional technology. In comparison, second generation biofuels are made from lignocellulosic biomass or woody crops, agricultural residues or waste, which makes it harder to extract the required fuel. A series of physical and chemical treatments might be required to convert lignocellulosic biomass to liquid fuels suitable for transportation.

Sustainable Biofuels

Biofuels in the form of liquid fuels derived from plant materials are entering the market, driven mainly by the perception that they reduce climate gas emissions, and also by factors such as oil price spikes and the need for increased energy security. However, many of the biofuels that are currently being supplied have been criticised for their adverse impacts on the natural environment, food security, and land use. In 2008, the Nobel-prize winning chemist Paul J. Crutzen published findings that the release of nitrous oxide (N_2O) emissions in the production of biofuels means that overall they contribute more to global warming than the fossil fuels they replace.

The challenge is to support biofuel development, including the development of new cellulosic technologies, with responsible policies and economic instruments to help ensure that biofuel commercialization is sustainable. Responsible commercialization of biofuels represents an opportunity to enhance sustainable economic prospects in Africa, Latin America and Asia.

According to the Rocky Mountain Institute, sound biofuel production practices would not hamper food and fibre production, nor cause water or environmental problems, and would enhance soil fertility. The selection of land on which to grow the feedstocks is a critical component of the ability of biofuels to deliver sustainable solutions. A key consideration is the minimisation of biofuel competition for prime cropland.

Biofuels by Region

Bio Diesel Powered Fast Attack Craft Of Indian Navy patrolling during IFR 2016. The green bands on the vessels are indicative of the fact that the vessels are powered by bio-diesel

There are international organizations such as IEA Bioenergy, established in 1978 by the OECD International Energy Agency (IEA), with the aim of improving cooperation and information exchange between countries that have national programs in bioenergy research, development and deployment. The UN International Biofuels Forum is formed by Brazil, China, India, Pakistan, South Africa, the United States and the European Commission. The world leaders in biofuel development and use are Brazil, the United States, France, Sweden and Germany. Russia also has 22% of world's forest, and is a big biomass (solid biofuels) supplier. In 2010, Russian pulp and paper maker, Vyborgskaya Cellulose, said they would be producing pellets that can be used in heat and electricity generation from its plant in Vyborg by the end of the year. The plant will eventually produce about 900,000 tons of pellets per year, making it the largest in the world once operational.

Biofuels currently make up 3.1% of the total road transport fuel in the UK or 1,440 million litres. By 2020, 10% of the energy used in UK road and rail transport must come from renewable sources – this is the equivalent of replacing 4.3 million tonnes of fossil oil each year. Conventional biofuels are likely to produce between 3.7 and 6.6% of the energy needed in road and rail transport, while advanced biofuels could meet up to 4.3% of the UK's renewable transport fuel target by 2020.

Air Pollution

Biofuels are different from fossil fuels in regard to greenhouse gases but are similar to fossil fuels in that biofuels contribute to air pollution. Burning produces airborne carbon particulates, carbon monoxide and nitrous oxides. The WHO estimates 3.7 million premature deaths worldwide in 2012 due to air pollution. Brazil burns significant amounts of ethanol biofuel. Gas chromatograph studies were performed of ambient air in São Paulo, Brazil, and compared to Osaka, Japan, which does not burn ethanol fuel. Atmospheric Formaldehyde was 160% higher in Brazil, and Acetaldehyde was 260% higher.

Debates Regarding the Production and Use of Biofuel

There are various social, economic, environmental and technical issues with biofuel production and use, which have been discussed in the popular media and scientific journals. These include: the effect of moderating oil prices, the "food vs fuel" debate, food prices, poverty reduction potential, energy ratio, energy requirements, carbon emissions levels, sustainable biofuel production, deforestation and soil erosion, loss of biodiversity, impact on water resources, the possible modifications necessary to run the engine on biofuel, as well as energy balance and efficiency. The International Resource Panel, which provides independent scientific assessments and expert advice on a variety of resource-related themes, assessed the issues relating to biofuel use in its first report *Towards sustainable production and use of resources: Assessing Biofuels*. "Assessing Biofuels" outlined the wider and interrelated factors that need to be considered when deciding on the relative merits of pursuing one biofuel over another. It concluded that not all biofuels perform equally in terms of their impact on climate, energy security and ecosystems, and suggested that environmental and social impacts need to be assessed throughout the entire life-cycle.

Another issue with biofuel use and production is the US has changed mandates many times because the production has been taking longer than expected. The Renewable Fuel Standard (RFS) set by congress for 2010 was pushed back to at best 2012 to produce 100 million gallons of pure ethanol (not blended with a fossil fuel).

Current Research

Research is ongoing into finding more suitable biofuel crops and improving the oil yields of these crops. Using the current yields, vast amounts of land and fresh water would be needed to produce enough oil to completely replace fossil fuel usage. It would require twice the land area of the US to be devoted to soybean production, or two-thirds to be devoted to rapeseed production, to meet current US heating and transportation needs.

Specially bred mustard varieties can produce reasonably high oil yields and are very useful in crop rotation with cereals, and have the added benefit that the meal left over after the oil has been pressed out can act as an effective and biodegradable pesticide.

The NFESC, with Santa Barbara-based Biodiesel Industries, is working to develop biofuels technologies for the US navy and military, one of the largest diesel fuel users in the world. A group of Spanish developers working for a company called Ecofasa announced a new biofuel made from trash. The fuel is created from general urban waste which is treated by bacteria to produce fatty acids, which can be used to make biofuels.

Ethanol Biofuels

As the primary source of biofuels in North America, many organizations are conducting research in the area of ethanol production. The National Corn-to-Ethanol Research Center (NCERC) is a research division of Southern Illinois University Edwardsville dedicated solely to ethanol-based biofuel research projects. On the federal level, the USDA conducts a large amount of research regarding ethanol production in the United States. Much of this research is targeted toward the effect of ethanol production on domestic food markets. A division of the U.S. Department of Energy, the National Renewable Energy Laboratory (NREL), has also conducted various ethanol research projects, mainly in the area of cellulosic ethanol.

Cellulosic ethanol commercialization is the process of building an industry out of methods of turning cellulose-containing organic matter into fuel. Companies, such as Iogen, POET, and Abengoa, are building refineries that can process biomass and turn it into bioethanol. Companies, such as Diversa, Novozymes, and Dyadic, are producing enzymes that could enable a cellulosic ethanol future. The shift from food crop feedstocks to waste residues and native grasses offers significant opportunities for a range of players, from farmers to biotechnology firms, and from project developers to investors.

As of 2013, the first commercial-scale plants to produce cellulosic biofuels have begun operating. Multiple pathways for the conversion of different biofuel feedstocks are being used. In the next few years, the cost data of these technologies operating at commercial scale, and their relative performance, will become available. Lessons learnt will lower the costs of the industrial processes involved.

In parts of Asia and Africa where drylands prevail, sweet sorghum is being investigated as a potential source of food, feed and fuel combined. The crop is particularly suitable for growing in arid conditions, as it only extracts one seventh of the water used by sugarcane. In India, and other places, sweet sorghum stalks are used to produce biofuel by squeezing the juice and then fermenting into ethanol.

A study by researchers at the International Crops Research Institute for the Semi-Arid Tropics (ICRISAT) found that growing sweet sorghum instead of grain sorghum could increase farmers incomes by US$40 per hectare per crop because it can provide fuel in addition to food and animal feed. With grain sorghum currently grown on over 11 million hectares (ha) in Asia and on 23.4 million ha in Africa, a switch to sweet sorghum could have a considerable economic impact.

Algae Biofuels

From 1978 to 1996, the US NREL experimented with using algae as a biofuels source in the "Aquatic Species Program". A self-published article by Michael Briggs, at the UNH Biofuels Group, offers estimates for the realistic replacement of all vehicular fuel with biofuels by using algae that have a natural oil content greater than 50%, which Briggs suggests can be grown on algae ponds at wastewater treatment plants. This oil-rich algae can then be extracted from the system and processed into biofuels, with the dried remainder further reprocessed to create ethanol. The production of algae to harvest oil for biofuels has not yet been undertaken on a commercial scale, but feasibility studies have been conducted to arrive at the above yield estimate. In addition to its projected high yield, algaculture — unlike crop-based biofuels — does not entail a decrease in food production, since it requires neither farmland nor fresh water. Many companies are pursuing algae bioreactors for various purposes, including scaling up biofuels production to commercial levels. Prof. Rodrigo E. Teixeira from the University of Alabama in Huntsville demonstrated the extraction of biofuels lipids from wet algae using a simple and economical reaction in ionic liquids.

Jatropha

Several groups in various sectors are conducting research on *Jatropha curcas*, a poisonous shrub-like tree that produces seeds considered by many to be a viable source of biofuels feedstock oil. Much of this research focuses on improving the overall per acre oil yield of Jatropha through advancements in genetics, soil science, and horticultural practices.

SG Biofuels, a San Diego-based jatropha developer, has used molecular breeding and biotechnology to produce elite hybrid seeds that show significant yield improvements over first-generation varieties. SG Biofuels also claims additional benefits have arisen from such strains, including improved flowering synchronicity, higher resistance to pests and diseases, and increased cold-weather tolerance.

Plant Research International, a department of the Wageningen University and Research Centre in the Netherlands, maintains an ongoing Jatropha Evaluation Project that examines the feasibility of large-scale jatropha cultivation through field and laboratory experiments. The Center for Sustainable Energy Farming (CfSEF) is a Los Angeles-based nonprofit research organization dedicated to jatropha research in the areas of plant science, agronomy, and horticulture. Successful exploration of these disciplines is projected to increase jatropha farm production yields by 200-300% in the next 10 years.

Fungi

A group at the Russian Academy of Sciences in Moscow, in a 2008 paper, stated they had isolated large amounts of lipids from single-celled fungi and turned it into biofuels in an economically efficient manner. More research on this fungal species, *Cunninghamella japonica*, and others, is likely to appear in the near future. The recent discovery of a variant of the fungus *Gliocladium roseum* (later renamed Ascocoryne sarcoides) points toward the production of so-called myco-diesel from cellulose. This organism was recently discovered in the rainforests of northern Patagonia, and has the unique capability of converting cellulose into medium-length hydrocarbons typically found in diesel fuel. Many other fungi that can degrade cellulose and other polymers have been observed to produce molecules that are currently being engineered using organisms from other kingdoms, suggesting that fungi may play a large role in the bio-production of fuels in the future (reviewed in).

Animal Gut Bacteria

Microbial gastrointestinal flora in a variety of animals have shown potential for the production of biofuels. Recent research has shown that TU-103, a strain of *Clostridium* bacteria found in Zebra feces, can convert nearly any form of cellulose into butanol fuel. Microbes in panda waste are being investigated for their use in creating biofuels from bamboo and other plant materials. There has also been substantial research into the technology of using the gut microbiomes of wood-feeding insects for the conversion of lignocellulotic material into biofuel.

Greenhouse Gas Emissions

Some scientists have expressed concerns about land-use change in response to greater demand for crops to use for biofuel and the subsequent carbon emissions. The payback period, that is, the time it will take biofuels to pay back the carbon debt they acquire due to land-use change, has been estimated to be between 100 and 1000 years, depending on the specific instance and location of land-use change. However, no-till practices combined with cover-crop practices can reduce the payback period to three years for grassland conversion and 14 years for forest conversion.

A study conducted in the Tocantis State, in northern Brazil, found that many families were cutting down forests in order to produce two conglomerates of oilseed plants, the J. curcas (JC group) and the R. communis (RC group). This region is composed of 15% Amazonian rainforest with high biodiversity, and 80% Cerrado forest with lower biodiversity. During the study, the farmers that planted the JC group released over 2193 Mg CO_2, while losing 53-105 Mg CO_2 sequestration from deforestation; and the RC group farmers released 562 Mg CO_2, while losing 48-90 Mg CO_2 to be sequestered from forest depletion. The production of these types of biofuels not only led into an increased emission of carbon dioxide, but also to lower efficiency of forests to absorb the gases that these farms were emitting. This has to do with the amount of fossil fuel the production of fuel crops involves. In addition, the intensive use of monocropping agriculture requires large amounts of water irrigation, as well as of fertilizers, herbicides and pesticides. This does not only lead to poisonous chemicals to disperse on water runoff, but also to the emission of nitrous oxide (NO_2) as a fertilizer byproduct, which is three hundred times more efficient in producing a greenhouse effect than carbon dioxide (CO_2).

Converting rainforests, peatlands, savannas, or grasslands to produce food crop–based biofuels in Brazil, Southeast Asia, and the United States creates a "biofuel carbon debt" by releasing 17 to 420 times more CO_2 than the annual greenhouse gas (GHG) reductions that these biofuels would provide by displacing fossil fuels. Biofuels made from waste biomass or from biomass grown on abandoned agricultural lands incur little to no carbon debt.

Water Use

In addition to water required to grow crops, biofuel facilities require significant process water.

Natural Oil Polyols

Natural oil polyols, also known as NOPs or biopolyols, are polyols derived from vegetable oils by several different techniques. The primary use for these materials is in the production of polyurethanes. Most NOPs qualify as biobased products, as defined by the United States Secretary of Agriculture in the Farm Security and Rural Investment Act of 2002.

NOPs all have similar sources and applications, but the materials themselves can be quite different, depending on how they are made. All are clear liquids, ranging from colorless to medium yellow. Their viscosity is also variable and is usually a function of the molecular weight and the average number of hydroxyl groups per molecule (higher mw and higher hydroxyl content both giving higher viscosity.) Odor is a significant property which is different from NOP to NOP. Most NOPs are still quite similar chemically to their parent vegetable oils and as such are prone to becoming rancid. This involves autoxidation of fatty acid chains containing carbon-carbon double bonds and ultimately the formation of odoriferous, low molecular weight aldehydes, ketones and carboxylic acids. Odor is undesirable in the NOPs themselves, but more importantly, in the materials made from them.

There are a limited number of naturally occurring vegetable oils (triglycerides) which contain the unreacted hydroxyl groups that account for both the name and important reactivity of these polyols. Castor oil is the only commercially available natural oil polyol that is produced directly from a plant source: all other NOPs require chemical modification of the oils directly available from plants.

The hope is that using renewable resources as feedstocks for chemical processes will reduce the environmental footprint by reducing the demand on non-renewable fossil fuels currently used in the chemical industry and reduce the overall production of carbon dioxide, the most notable greenhouse gas. One NOP producer, Cargill, estimates that its BiOH(TM)polyol manufacturing process produces 36% less global warming emissions (carbon dioxide), a 61% reduction in non-renewable energy use (burning fossil fuels), and a 23% reduction in the total energy demand, all relative to polyols produced from petrochemicals.

Sources of Natural Oil Polyols

Ninety percent of the fatty acids that make up castor oil is ricinoleic acid, which has a hydroxyl group on C-12 and a carbon-carbon double bond. The structure below shows the major component of castor oil which is composed of the tri-ester of rincinoleic acid and glycerin:

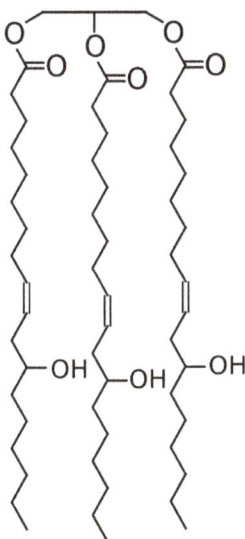

Other vegetable oils - such as soy bean oil, peanut oil, and canola oil - contain carbon-carbon double bonds, but no hydroxyl groups. There are several processes used to introduce hydroxyl groups onto the carbon chain of the fatty acids, and most of these involve oxidation of the C-C double bond. Treatment of the vegetal oils with ozone cleaves the double bond, and esters or alcohols can be made, depending on the conditions used to process the ozonolysis product. The example below shows the reaction of triolein with ozone and ethylene glycol.

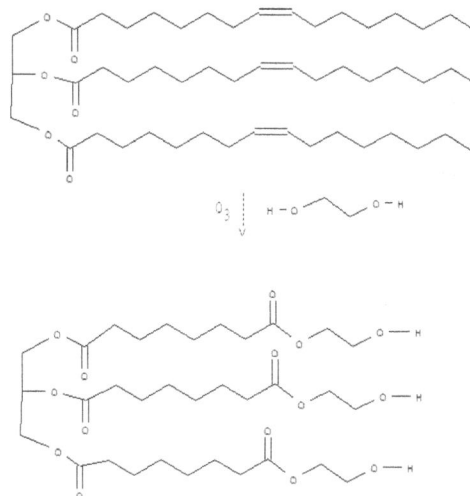

Air oxidation, (autoxidation), the chemistry involved in the "drying" of drying oils, gives increased molecular weight and introduces hydroxyl groups. The radical reactions involved in autoxidation can produce a complex mixture of crosslinked and oxidized triglycerides. Treatment of vegetable oils with peroxy acids gives epoxides which can be reacted with nucleophiles to give hydroxyl groups. This can be done as a one-step process. Note that in the example shown below only one of the three fatty acid chains is drawn fully, the other part of the molecule is represented by "R_1" and the nucleophile is unspecified. Earlier examples also include acid catalyzed ring opening of epoxidized soybean oil to make oleochemical polyols for polyurethane foams and acid catalyzed ring opening of soy fatty acid methyl esters with multifunctional polyols to form new polyols for casting resins.

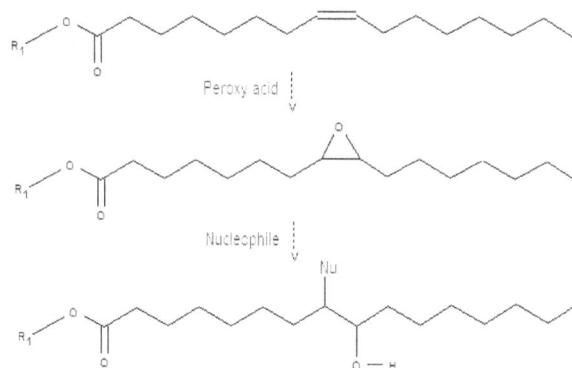

Triglycerides of unsaturated (containing carbon-carbon double bonds) fatty acids or methyl esters of these acids, can be treated with carbon monoxide and hydrogen in the presence of a metal catalyst to add a -CHO (formyl) groups to the chain (hydroformylation reaction) followed by hydrogenation to give the needed hydroxyl groups. In this case R_1 can be the rest of the triglyceride, or a smaller group such as methyl (in which case the substrate would be similar to biodiesel). If R=Me then additional reactions like transesterification are needed to build up a polyol.

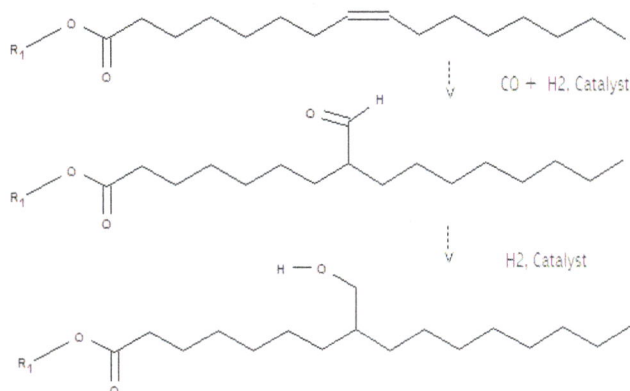

Polyols from non conventional vegetable oils - neem oil, karanja oil, linseed oil, cottonseed oil, etc. - are potential industrial feedstocks for polyurethanes.

Uses

Castor oil has found numerous applications, many of them due to the presence of the hydroxyl group that allows chemical derivatization of the oil or modifies the properties of castor oil relative to vegetable oils which do not have the hydroxyl group. Castor oil undergoes most of the reactions that alcohols do, but the most industrially important one is reaction with diisocyanates to make polyurethanes.

Castor oil by itself has been used in making a variety of polyurethane products, ranging from coatings to foams, and the use of castor oil derivatives continues to be an area of active development. Castor oil derivatized with propylene oxide makes polyurethane foam for mattresses and yet another new derivative is used in coatings

Apart from castor oil, which is a relatively expensive vegetable oil and is not produced domestically in many industrialized countries, the use of polyols derived from vegetable oils to make polyurethane products began attracting attention beginning around 2004. The rising costs of petrochemical feedstocks and an enhanced public desire for environmentally friendly green products have created a demand for these materials. One of the most vocal supporters of these polyurethanes made using natural oil polyols is the Ford Motor Company, which debuted polyurethane foam made using soy oil in the seats of its 2008 Ford Mustang. Ford has since placed soy foam seating in all its North American vehicle platforms. The interest of automakers is responsible for much of the work being done on the use of NOPs in polyurethane products for use in cars, for example is seats, and headrests, armrests, soundproofing, and even body panels.

One of the first uses for NOPs (other than castor oil) was to make spray-on polyurethane foam insulation for buildings.

NOPs are also finding use in polyurethane slab foam used to make conventional mattresses as well as memory foam mattresses.

The characteristics of NOPs can be varied over a very wide range. This can be done by selection of the base Natural Oil (or oils) used to make up the NOP. Also, using known and increasingly novel (Garrett & Du) chemical techniques, it is possible to graft additional groups onto the triglyceride chains of the NOP and change its processing characteristics and this in turn will change and modify in a controlled manner, the physical properties of the final article which the NOP is being used to produce. Differences and modifications in the process regime and reaction conditions used to make a given NOP also generally lead to different chemical architectures and therefore different end use performance of that NOP; so that even though two NOPs may have been made from the same Natural Oil root, they may be surprisingly different when used and, will produce a detectably different end product too. Commercially, (since 2012) NOPs are available and made from; sawgrass oil, soybean oil, castor oil (as an grafted NOP), rapeseed oil, palm oil (kernel and mesocarp), and coconut oil. There is also some work being done on NOPs made from Natural Animal oils.

Initially in the US, and since early 2010, it has been routinely possible to replace over 50% of petrochemical-based polyols with NOPs for use in slab foams sold into the mass market, furniture and bedding industries. The commercialised technology also eliminates or greatly reduces the odor problem, mentioned above, normally associated with the use of NOPs. This is particularly important when the NOP is to be used at ever higher percentage levels, to try to reduce dependency on petrochemical materials, and to produce materials for use in the domestic and contract furniture segments which are historically very sensitive to "chemical" odors in the final foam product in people's homes and places of work.

Amongst other useful effects of using high levels of Natural Oil Polyols to make foams are the improvements seen in the long-term performance of the foam under humid conditions and also on the flammability of the foams; compared to equivalent foams made without the presence of the NOP. People perspire; and so foams used for the construction of matrasses or furniture will, over time, tend to feel softer and give less support. The perspiration gradually softens the foam. Foams made with high levels of NOPs are much less prone to this problem, so that the useful lifetime of the upholstered product can be extended. The use of high levels of NOP also make it possible to manufacture foams with flame retardants which are permanent, and therefore are not later emitted into the household or work place environment. These relatively recently developed materials can be added at very low levels to NOP foams to pass such well known tests as California Technical Bulletin 117, which is a well-known flammability test for furniture. These permanent flame-retardants are halogen free and key into the foam matrix and are therefore fixed there. An additional effect of using these new, highly efficient, permanent flame retardants, is that the smoke seen during these standard fire tests, may be considerably reduced compared to that produced when testing foams made using non-permanent flame retardant materials, which do not key themselves into the foam structure. More recent work during 2014 with this "Green Chemistry" has shown that foams containing about 50 percent by weight of natural oils can be made which produce far less smoke when involved in fire situations. The ability of these low emission foams to reduce smoke emissions by up to 80% is an interesting property which will aid escape from fire situations and also lessen the risks for first responders i.e. emergency services in general and fire department personnel in particular.

Other technology can be combined with these flammability characteristics to give foams, which have extremely low overall emissions of volatile organic compounds, known as VOCs.

Bio-based Material

A bio-based material is a material intentionally made from substances derived from living (or once-living) organisms. These materials are sometimes referred to as biomaterials, but this word also has another meaning. Strictly the definition could include many common materials such as wood and leather, but it typically refers to modern materials that have undergone more extensive processing. Unprocessed materials may be called biotic material. Bio-based materials or biomaterials fall under the broader category of bioproducts or bio-based products which includes materials, chemicals and energy derived from renewable biological resources.

Bio-based materials are often biodegradable, but this is not always the case.

Examples include:

- cellulose fibers — fibers made from reconstituted cellulose.

- casein — a phosphoprotein extracted from milk during the process of creating low fat milk, it is processed in various ways to make: plastic, dietary supplements for body builders, glue, cotton candy, protective coatings, paints, and occurs naturally in cheese, giving it a creamy texture.

- polylactic acid — a polymer produced by industrial fermentation

- bioplastics — include a soy oil based plastic now being used to make body panels for John Deere tractors

- engineered wood — products such as oriented strand board and particle board

- zein — a natural biopolymer which is the most abundant corn protein

- cornstarch — the starch of the maize grain, used to make packing pellets

- grease — lubricants made from vegetable oils, including soybean oil, that can replace petroleum based lubricants

Polylactic Acid

Poly(lactic acid) or polylactic acid or polylactide (PLA) is a biodegradable and bioactive thermoplastic aliphatic polyester derived from renewable resources, such as corn starch (in the United States and Canada), tapioca roots, chips or starch (mostly in Asia), or sugarcane (in the rest of the world). In 2010, PLA had the second highest consumption volume of any bioplastic of the world.

The name "polylactic acid" does not comply with IUPAC standard nomenclature, and is potentially ambiguous or confusing, because PLA is not a polyacid (polyelectrolyte), but rather a polyester.

Production

Producers have several industrial routes to usable (i.e. high molecular weight) PLA. Two main monomers are used: lactic acid, and the cyclic di-ester, lactide. The most common route to PLA is the ring-opening polymerization of lactide with various metal catalysts (typically tin octoate) in solution, in the melt, or as a suspension. The metal-catalyzed reaction tends to cause racemization of the PLA, reducing its stereoregularity compared to the starting material (usually corn starch).

Another route to PLA is the direct condensation of lactic acid monomers. This process needs to be carried out at less than 200 °C; above that temperature, the entropically favored lactide monomer is generated. This reaction generates one equivalent of water for every condensation (esterification) step, and that is undesirable because water causes chain-transfer leading to low molecular weight material. The direct condensation is thus performed in a stepwise fashion, where lactic acid is first oligomerized to PLA oligomers. Thereafter, polycondensation is done in the melt or as a solution, where short oligomeric units are combined to give a high molecular weight polymer strand. Water removal by application of a vacuum or by azeotropic distillation is crucial to favor polycondensation over transesterification. Molecular weights of 130 kDa can be obtained this way. Even higher molecular weights can be attained by carefully crystallizing the crude polymer from the melt. Carboxylic acid and alcohol end groups are thus concentrated in the amorphous region of the solid polymer, and so they can react. Molecular weights of 128–152 kDa are obtainable thus.

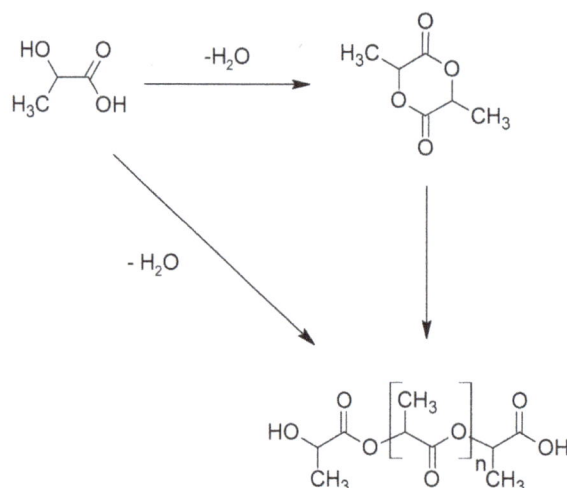

Polymerization of a racemic mixture of L- and D-lactides usually leads to the synthesis of poly-DL-lactide (PDLLA), which is amorphous. Use of stereospecific catalysts can lead to heterotactic PLA which has been found to show crystallinity. The degree of crystallinity, and hence many important properties, is largely controlled by the ratio of D to L enantiomers used, and to a lesser extent on the type of catalyst used. Apart from lactic acid and lactide, lactic acid O-carboxyanhydride ("lac-OCA"), a five-membered cyclic compound has been used academically as well. This compound is more reactive than lactide, because its polymerization is driven by the loss of one equivalent of carbon dioxide per equivalent of lactic acid. Water is not a co-product.

The direct biosynthesis of PLA similar to the poly(hydroxyalkanoate)s has been reported as well.

Manufacturers

As of June 2010, NatureWorks was the primary producer of PLA (bioplastic) in the United States. Other companies involved in PLA manufacturing are Evonik Industries (Germany), Corbion PURAC Biomaterials (The Netherlands) who have announced a new 75,000 ton PLA plant in Thailand by 2018, and several Chinese manufacturers. The primary producer of PDLLA is Evonik Industries and Corbion PURAC. Evonik Industries is a specialty chemical company that is industry leading in approximately 80% of the markets they participate. The Resomer brand of PDLLA is produced in the Health and Nutrition business segment. Corbion PURAC is a listed company in the Netherlands, and operating plants worldwide, and the only producer of PDLA, produced from the D-isomer of lactid acid. Galactic and Total Petrochemicals operate a joint venture, Futerro, which is developing a second generation polylactic acid product. This project includes the building of a PLA pilot plant in Belgium capable of producing 1,500 tonnes/year.

Chemical and Physical Properties

Due to the chiral nature of lactic acid, several distinct forms of polylactide exist: poly-L-lactide (PLLA) is the product resulting from polymerization of L,L-lactide (also known as L-lactide). PLLA has a crystallinity of around 37%, a glass transition temperature 60–65 °C, a melting temperature 173–178 °C and a tensile modulus 2.7–16 GPa. Heat-resistant PLA can withstand temperatures of 110 °C. PLA is soluble in chlorinated solvents, hot benzene, tetrahydrofuran, and dioxane.

Polylactic acid can be processed like most thermoplastics into fiber (for example, using conventional melt spinning processes) and film. PLA has similar mechanical properties to PETE polymer, but has a significantly lower maximum continuous use temperature. The tensile strength for 3-D printed PLA was previously determined. It was found to range widely depending on printing conditions, which were obtained using RepRap 3-D printers. Results of a recent study gave a printed tensile strength of around 50 MPa and show that the act of 3-D printing PLA affects its properties—they showed a strong relationship between tensile strength and percent crystallinity of a 3-D printed sample and a strong relationship between percent crystallinity and the extruder temperature.

The melting temperature of PLLA can be increased by 40–50 °C and its heat deflection temperature can be increased from approximately 60 °C to up to 190 °C by physically blending the polymer with PDLA (poly-D-lactide). PDLA and PLLA form a highly regular stereocomplex with increased crystallinity. The temperature stability is maximised when a 1:1 blend is used, but even at lower concentrations of 3–10% of PDLA, there is still a substantial improvement. In the latter case, PDLA acts as a nucleating agent, thereby increasing the crystallization rate. Biodegradation of PDLA is slower than for PLA due to the higher crystallinity of PDLA.

There is also poly(L-lactide-*co*-D,L-lactide) (PLDLLA) – used as PLDLLA/TCP scaffolds for bone engineering.

Applications

Mulch film made of PLA-blend "bioflex"

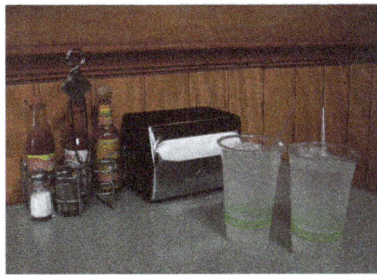

Biodegradable PLA cups in use at a restaurant

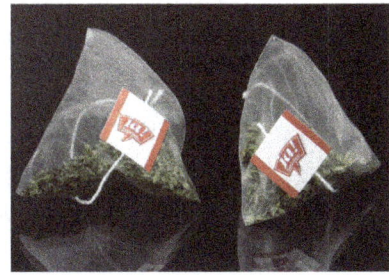

Tea bags made of PLA. Peppermint tea is enclosed.

PLA can be processed by extrusion such as 3d printing, injection molding, film and sheet casting, and spinning, providing access to a wide range of materials.

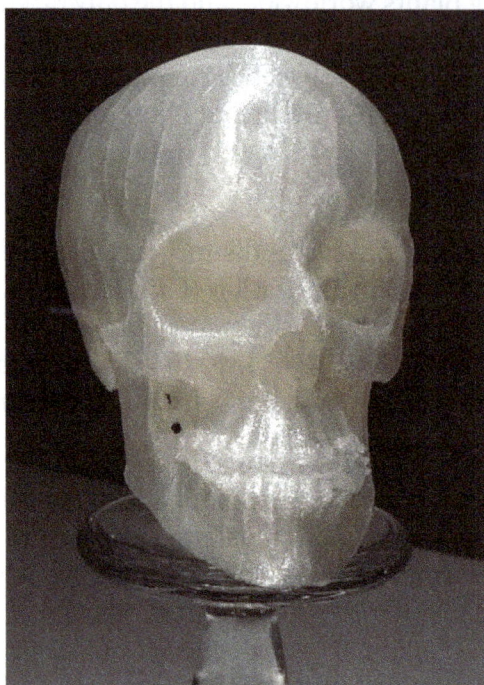

3D Printed Human skull with data from Computed Tomography. Transparent PLA.

PLA is used as a feedstock material in desktop fused filament fabrication-based 3D printers (e.g. RepRap). PLA printed solids can be encased in plaster-like moulding materials, then burned out in a furnace, so that the resulting void can be filled with molten metal. This is known as "lost PLA casting", a type of investment casting.

Being able to degrade into innocuous lactic acid, PLA is used as medical implants in the form of anchors, screws, plates, pins, rods, and as a mesh. Depending on the exact type used, it breaks down inside the body within 6 months to 2 years. This gradual degradation is desirable for a support structure, because it gradually transfers the load to the body (e.g. the bone) as that area heals. The strength characteristics of PLA and PLLA implants are well documented.

PLA can also be used as a decomposable packaging material, either cast, injection-molded, or spun. Cups and bags have been made from this material. In the form of a film, it shrinks upon heating, allowing it to be used in shrink tunnels. It is useful for producing loose-fill packaging, compost bags, food packaging, and disposable tableware. In the form of fibers and nonwoven fabrics, PLA also has many potential uses, for example as upholstery, disposable garments, awnings, feminine hygiene products, and diapers.

Racemic and regular PLLA has a low glass transition temperature, which is undesirable. A stereocomplex of PDLA and PLLA has a higher glass transition temperatures, lending it more mechanical strength. It has a wide range of applications, such as woven shirts (ironability), microwavable trays, hot-fill applications and even engineering plastics (in this case, the stereocomplex is blended with a rubber-like polymer such as ABS). Such blends also have good form stability and visual transparency, making them useful for low-end packaging applications. Pure poly-L-lactic acid (PLLA), on the other hand, is the main ingredient in Sculptra, a long-lasting facial volume enhancer, primarily used for lipoatrophy of cheeks. Progress in biotechnology has resulted in the development of commercial production of the D enantiomer form, something that was not possible until recently.

Recycling

PLA has SPI resin ID code 7

Currently, the SPI resin identification code 7 ("others") is applicable for PLA. In Belgium, Galactic started the first pilot unit to chemically recycle PLA (Loopla). Unlike mechanical recycling, waste material can hold various contaminants. Polylactic acid can be recycled to monomer by thermal depolymerization or hydrolysis. When purified, the monomer can be used for the manufacturing of virgin PLA with no loss of original properties (cradle-to-cradle recycling).

Degradation

Amycolatopsis and *Saccharotrix* are able to degrade PLA. A purified protease from *Amycolatopsis* sp., PLA depolymerase, can also degrade PLA. Enzymes such as pronase and most effectively proteinase K from *Tritirachium album* degrade PLA.

Pure PLLA foams undergo selective hydrolysis when placed in an environment of Dulbecco's modified Eagle's medium (DMEM) supplemented with fetal bovine serum (FBS) (a solution mimicking body fluid). After 30 days of submersion in DMEM+FBS, a PLLA scaffold lost about 20% of its weight.

Bioplastic

Bioplastics are plastics derived from renewable biomass sources, such as vegetable fats and oils, corn starch, or microbiota. Bioplastic can be made from agricultural by-products and also from used plastic bottles and other containers using microorganisms. Common plastics, such as fossil-fuel plastics (also called petrobased polymers), are derived from petroleum or natural gas. Production of such plastics tends to require more fossil fuels and to produce more greenhouse gases than the production of biobased polymers (bioplastics). Some, but not all, bioplastics are designed to biodegrade. Biodegradable bioplastics can break down in either anaerobic or aerobic environments, depending on how they are manufactured. Bioplastics can be composed of starches, cellulose, biopolymers, and a variety of other materials.

IUPAC Definition:

Biobased polymer derived from the *biomass* or issued from monomers derived from the biomass and which, at some stage in its processing into finished products, can be shaped by flow.

Note 1: Bioplastic is generally used as the opposite of polymer derived from fossil resources.

Note 2: Bioplastic is misleading because it suggests that any polymer derived from the biomass is *environmentally friendly*.

Note 3: The use of the term "bioplastic" is discouraged. Use the expression "biobased polymer".

Note 4: A biobased polymer similar to a petrobased one does not imply any superiority with respect to the environment unless the comparison of respective *life cycle assessments* is favourable.

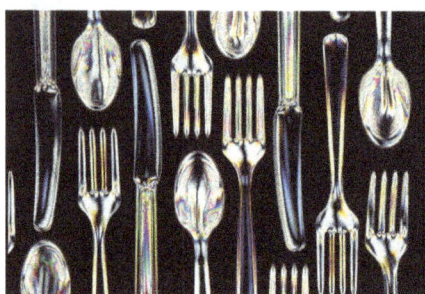

Biodegradable plastic utensils

Packaging peanuts made from bioplastics (thermoplastic starch)

Plastics packaging made from bioplastics and other biodegradable plastics

Applications

Bioplastics are used for disposable items, such as packaging, crockery, cutlery, pots, bowls, and straws. They are also often used for bags, trays, fruit and vegetable containers and blister foils, egg cartons, meat packaging, vegetables, and bottling for soft drinks and dairy products.

Flower wrapping made of PLA-blend bio-flex

These plastics are also used in non-disposable applications including mobile phone casings, carpet fibers, insulation car interiors, fuel lines, and plastic piping. New electroactive bioplastics are being developed that can be used to carry electric current. In these areas, the goal is not biodegradability, but to create items from sustainable resources.

Medical implants made of PLA (polylactic acid), which dissolve in the body, can save patients a second operation. Compostable mulch films can also be produced from starch polymers and used in agriculture. These films do not have to be collected after use on farm fields.

Biopolymers are available as coatings for paper rather than the more common petrochemical coatings.

Bioplastic Types

Starch-based Plastics

Thermoplastic starch currently represents the most widely used bioplastic, constituting about 50 percent of the bioplastics market. Simple starch bioplastic can be made at home. Pure starch is able to absorb humidity, and is thus a suitable material for the production of drug capsules by the pharmaceutical sector. Flexibiliser and plasticiser such as sorbitol and glycerine can also be added so the starch can also be processed thermo-plastically. The characteristics of the resulting bioplastic (also called "thermo-plastical starch") can be tailored to specific needs by adjusting the amounts of these additives.

Starch-based bioplastics are often blended with biodegradable polyesters to produce starch / polycaprolactone or starch/Ecoflex (polybutylene adipate-co-terephthalate produced by BASF). blends. These blends are used for industrial applications and are also compostable. Other producers, such as Roquette, have developed other starch/polyolefin blends. These blends are not biodegradable, but have a lower carbon footprint than petroleum-based plastics used for the same applications.

Cellulose-Based Plastics

A packaging blister made from cellulose acetate, a bioplastic

Cellulose bioplastics are mainly the cellulose esters, (including cellulose acetate and nitrocellulose) and their derivatives, including celluloid.

Protein-based Plastics

Bioplastics can be made from proteins from different sources. For example wheat gluten and casein show promising properties as a raw material for different biodegradable polymers.

Some Aliphatic Polyesters

The aliphatic biopolyesters are mainly polyhydroxyalkanoates (PHAs) like the poly-3-hydroxybutyrate (PHB), polyhydroxyvalerate (PHV) and polyhydroxyhexanoate (PHH).

Polylactic Acid (PLA)

Polylactic acid (PLA) is a transparent plastic produced from corn or dextrose. Its characteristics are similar to conventional petrochemical-based mass plastics (like PET, PS or PE), and it can be processed using standard equipment that already exists for the production of some conventional plastics. PLA and PLA blends generally come in the form of granulates with various properties, and are used in the plastic processing industry for the production of films, fibers, plastic containers, cups and bottles.

Poly-3-hydroxybutyrate (PHB)

The biopolymer poly-3-hydroxybutyrate (PHB) is a polyester produced by certain bacteria processing glucose, corn starch or wastewater. Its characteristics are similar to those of the petroplastic polypropylene. PHB production is increasing. The South American sugar industry, for example, has decided to expand PHB production to an industrial scale. PHB is distinguished primarily by its physical characteristics. It can be processed into a transparent film with a melting point higher than 130 degrees Celsius, and is biodegradable without residue.

Polyhydroxyalkanoates (PHA)

Polyhydroxyalkanoates are linear polyesters produced in nature by bacterial fermentation of sugar or lipids. They are produced by the bacteria to store carbon and energy. In industrial production, the polyester is extracted and purified from the bacteria by optimizing the conditions for the fermentation of sugar. More than 150 different monomers can be combined within this family to give materials with extremely different properties. PHA is more ductile and less elastic than other plastics, and it is also biodegradable. These plastics are being widely used in the medical industry.

Polyamide 11 (PA 11)

PA 11 is a biopolymer derived from natural oil. It is also known under the tradename Rilsan B, commercialized by Arkema. PA 11 belongs to the technical polymers family and is not biodegradable. Its properties are similar to those of PA 12, although emissions of greenhouse gases and consumption of nonrenewable resources are reduced during its production. Its thermal resistance is also superior to that of PA 12. It is used in high-performance applications like automotive fuel lines, pneumatic airbrake tubing, electrical cable antitermite sheathing, flexible oil and gas pipes, control fluid umbilicals, sports shoes, electronic device components, and catheters.

A similar plastic is Polyamide 410 (PA 410), derived 70% from castor oil, under the trade name EcoPaXX, commercialized by DSM. PA 410 is a high-performance polyamide that combines the benefits of a high melting point (approx. 250 °C), low moisture absorption and excellent resistance to various chemical substances.

Bio-derived Polyethylene

The basic building block (monomer) of polyethylene is ethylene. Ethylene is chemically similar to, and can be derived from ethanol, which can be produced by fermentation of agricultural feedstocks such as sugar cane or corn. Bio-derived polyethylene is chemically and physically identical to traditional polyethylene – it does not biodegrade but can be recycled. Bio derivation of polyethylene can also reduce greenhouse gas emissions considerably. Brazilian chemicals group Braskem claims that using its method of producing polyethylene from sugar cane ethanol captures (removes from the environment) 2.15 tonnes of CO_2 per tonne of Green Polyethylene produced.

Green PE has the same properties, performance and application versatility as fossil-based polyethylene, which makes it a drop-in replacement in the plastic production chain. For these same reasons, it is also recyclable in the same recycling chain used by traditional polyethylene. Because it is part of the portfolio of high-density polyethylene (HDPE) and linear low-density polyethylene (LLDPE) products, Green PE is an option for applications in rigid and flexible packaging, closures, bags and other products.

Genetically Modified Bioplastics

Genetic modification (GM) is also a challenge for the bioplastics industry. None of the currently available bioplastics – which can be considered first generation products – require the use of GM crops, although GM corn is the standard feedstock.

Looking further ahead, some of the second generation bioplastics manufacturing technologies under development employ the "plant factory" model, using genetically modified crops or genetically modified bacteria to optimise efficiency.

Polyhydroxyurethanes

Recently, there have been a large emphasis on producing biobased and isocyanate-free polyurethanes. One such example utilizes a spontaneous reaction between polyamines and cyclic carbonates to produce polyhydroxurethanes. Unlike traditional cross-linked polyurethanes, cross-linked polyhydroxyurethanes have been shown to be capable of recycling and reprocessing through dynamic transcarbamoylation reactions.

Environmental Impact

The environmental impact of bioplastics is often debated, as there are many different metrics for "greenness" (e.g., water use, energy use, deforestation, biodegradation, etc.) and tradeoffs often exist. The debate is also complicated by the fact that many different types of bioplastics exist, each with different environmental strengths and weaknesses, so not all bioplastics can be treated as equal.

Confectionery packaging made of PLA-blend bio-flex

Drinking straws made of PLA-blend bio-flex

Jar made of PLA-blend bio-flex, a bioplastic

The production and use of bioplastics is sometimes regarded as a more sustainable activity when compared with plastic production from petroleum (petroplastic), because it requires less fossil fuel for its production and also introduces fewer, net-new greenhouse emissions if it biodegrades. The use of bioplastics can also result in less hazardous waste than oil-derived plastics, which remain solid for hundreds of years.

Petroleum is often still used as a source of materials and energy in the production of bioplastic. Petroleum is required to power farm machinery, to irrigate crops, to produce fertilisers and pesticides, to transport crops and crop products to processing plants, to process raw materials, and ultimately to produce the bioplastic. However, it is possible to produce bioplastic using renewable energy sources and avoid the use of petroleum.

Italian bioplastic manufacturer Novamont states in its own environmental audit that producing one kilogram of its starch-based product uses 500 g of petroleum and consumes almost 80% of the energy required to produce a traditional polyethylene polymer. Environmental data from Nature-Works, the only commercial manufacturer of PLA (polylactic acid) bioplastic, says that making its plastic material delivers a fossil fuel saving of between 25 and 68 per cent compared with polyethylene, in part due to its purchasing of renewable energy certificates for its manufacturing plant.

A detailed study examining the process of manufacturing a number of common packaging items from traditional plastics and polylactic acid carried out by Franklin Associates and published by the Athena Institute shows that using bioplastic has a lower environmental impact for some products, and a higher environmental impact for others. This study, however, does not factor in the end-of-life environmental impact of these products, including possible methane emissions from landfills due to biodegradable plastics.

While production of most bioplastics results in reduced carbon dioxide emissions compared to traditional alternatives, there is concern that the creation of a global bioeconomy required to produce bioplastic in large quantities could contribute to an accelerated rate of deforestation and soil erosion, and could adversely affect water supplies. Careful management of a global bioeconomy would be required.

Other studies showed that bioplastics result in a 42% reduction in carbon footprint.

On October 21, 2010, a group of scientists reported that corn-based plastic ranked higher in environmental defects than the main products it replaces, such as HDPE, LDPE and PP. In the study, the production of corn-based plastics created more acidification, carcinogens, ecotoxicity, eutrophication, ozone depletion, respiratory effects and smog than the synthetic-based plastics they replaced. However the study also concluded that biopolymers trumped the other plastics for biodegradability, low toxicity, and use of renewable resources.

The American Carbon Registry has also released reports of nitrous oxide caused from corn growing which is 310 times more potent than CO_2. Pesticides are also used in growing corn-based plastic.

Bioplastics and Biodegradation

The terminology used in the bioplastics sector is sometimes misleading. Most in the industry use the term bioplastic to mean a plastic produced from a biological source. All (bio- and petroleum-based) plastics are technically biodegradable, meaning they can be degraded by microbes under suitable conditions. However, many degrade so slowly that they are considered non-biodegradable. Some petrochemical-based plastics are considered biodegradable, and may be used as an additive to improve the performance of commercial bioplastics. Non-biodegradable bioplastics are referred to as durable. The biodegradability of bioplastics depends on temperature, polymer stability, and available oxygen content. The European standard EN13432, published by the International Organization for Standardization, defines how quickly and to what extent a plastic must be degraded under the tightly controlled and aggressive conditions (at or above 140 °F) of an industrial composting unit for it to be considered biodegradable. This standard is recognized in many countries, including all of Europe, Japan and the US. However, it applies only to industrial composting units and does not set out a standard for home composting. Most bioplastics (e.g. PH) only biodegrade quickly in industrial composting units. These materials do not biodegrade quickly in ordinary compost piles or in the soil/water. Starch-based bioplastics are an exception, and will biodegrade in normal composting conditions.

Packaging air pillow made of PLA-blend bio-flex

The term "biodegradable plastic" has also been used by producers of specially modified petrochemical-based plastics that appear to biodegrade. Biodegradable plastic bag manufacturers that have misrepresented their product's biodegradability may now face legal action in the US state of California for the misleading use of the terms biodegradable or compostable. Traditional plastics such as polyethylene are degraded by ultra-violet (UV) light and oxygen. To prevent this, process manufacturers add stabilising chemicals. However, with the addition of a degradation initiator to the plastic, it is possible to achieve a controlled UV/oxidation disintegration process. This type of plastic may be referred to as *degradable plastic* or *oxy-degradable plastic* or *photodegradable plastic* because the process is not initiated by microbial action. While some degradable plastics manufacturers argue that degraded plastic residue will be attacked by microbes, these degradable materials do not meet the requirements of the EN13432 commercial composting standard. The bioplastics industry has widely criticized oxo-biodegradable plastics, which the industry association says do not meet its requirements. Oxo-biodegradable plastics – known as "oxos" – are conventional petroleum-based products with some additives that initiate degradation. The ASTM standard for oxo-biodegradables is called the Standard Guide for Exposing and Testing Plastics that Degrade in the Environment by a Combination of Oxidation and Biodegradation (ASTM 6954). Both EN 13432 and ASTM 6400 are specifically designed for PLA and Starch based products and should not be used as a guide for oxos.

Market

Prism pencil sharpener made from cellulose acetate biograde

Because of the fragmentation in the market and ambiguous definitions it is difficult to describe the total market size for bioplastics, but estimates put global production capacity at 327,000 tonnes. In contrast, global consumption of all flexible packaging is estimated at around 12.3 million tonnes.

COPA (Committee of Agricultural Organisation in the European Union) and COGEGA (General Committee for the Agricultural Cooperation in the European Union) have made an assessment of the potential of bioplastics in different sectors of the European economy:

Catering products: 450,000 tonnes per year

Organic waste bags: 100,000 tonnes per year

Biodegradable mulch foils: 130,000 tonnes per year

Biodegradable foils for diapers 80,000 tonnes per year

Diapers, 100% biodegradable: 240,000 tonnes per year

Foil packaging: 400,000 tonnes per year

Vegetable packaging: 400,000 tonnes per year

Tyre components: 200,000 tonnes per year

Total: 2,000,000 tonnes per year

In the years 2000 to 2008, worldwide consumption of biodegradable plastics based on starch, sugar, and cellulose – so far the three most important raw materials – has increased by 600%. The NNFCC predicted global annual capacity would grow more than six-fold to 2.1 million tonnes by 2013. BCC Research forecasts the global market for biodegradable polymers to grow at a compound average growth rate of more than 17 percent through 2012. Even so, bioplastics will encompass a small niche of the overall plastic market, which is forecast to reach 500 billion pounds (220 million tonnes) globally by 2010. Ceresana forecasts the world market for bioplastics to reach 5.8 billion US dollars in 2021 - i.e. three times more than in 2014.

Cost

At one time bioplastics were too expensive for consideration as a replacement for petroleum-based plastics. However, the lower temperatures needed to process bioplastics and the more stable supply of biomass combined with the increasing cost of crude oil make bioplastics' prices more competitive with regular plastics.

Research and Development

Bioplastics Development Center - University of Massachusetts Lowell

A pen made with bioplastics (Polylactide, PLA)

- In the early 1950s, amylomaize (>50% amylose content corn) was successfully bred and commercial bioplastics applications started to be explored.

- In 2004, NEC developed a flame retardant plastic, polylactic acid, without using toxic chemicals such as halogens and phosphorus compounds.

- In 2005, Fujitsu became one of the first technology companies to make personal computer cases from bioplastics, which are featured in their FMV-BIBLO NB80K line. Later, the French company Ashelvea (also listed on EU Energy Star registered partners), launched its fully recyclable PC with biodegradable plastic case "Evolutis", reported in "People Inspiring Philips", a series of 3 mini-documentaries to inspire Philips employees with some examples from the civil society.

- In 2007 Braskem of Brazil announced it had developed a route to manufacture high-density polyethylene (HDPE) using ethylene derived from sugar cane.

- In 2008, a University of Warwick team created a soap-free emulsion polymerization process which makes colloid particles of polymer dispersed in water, and in a one step process adds nanometre sized silica-based particles to the mix. The newly developed technology might be most applicable to multi-layered biodegradable packaging, which could gain more robustness and water barrier characteristics through the addition of a nano-particle coating.

Testing Procedures

A bioplastic shampoo bottle made of PLA-blend bio-flex

Industrial Compostability – EN 13432, ASTM D6400

The EN 13432 industrial standard is arguably the most international in scope. This standard must be met in order to claim that a plastic product is compostable in the European marketplace. In summary, it requires biodegradation of 90% of the materials in a lab within 90 days. The ASTM 6400 standard is the regulatory framework for the United States and sets a less stringent threshold of 60% biodegradation within 180 days for non-homopolymers, and 90% biodegradation of homopolymers within industrial composting conditions (temperatures at or above 140F). Municipal compost facilities do not see above 130F.

Many starch-based plastics, PLA-based plastics and certain aliphatic-aromatic co-polyester compounds, such as succinates and adipates, have obtained these certificates. Additive-based bioplastics sold as photodegradable or Oxo Biodegradable do not comply with these standards in their current form.

Compostability – ASTM D6002

The ASTM D 6002 method for determining the compostability of a plastic defined the word *compostable* as follows:

that which is capable of undergoing biological decomposition in a compost site such that the material is not visually distinguishable and breaks down into carbon dioxide, water, inorganic compounds and biomass at a rate consistent with known compostable materials.

This definition drew much criticism because, contrary to the way the word is traditionally defined, it completely divorces the process of "composting" from the necessity of it leading to humus/compost as the end product. The only criterion this standard *does* describe is that a compostable plastic must look to be going away as fast as something else one has already established to be compostable under the *traditional* definition.

Withdrawal of ASTM D 6002

In January 2011, the ASTM withdrew standard ASTM D 6002, which had provided plastic man-ufacturers with the legal credibility to label a plastic as compostable. Its description is as follows:

This guide covered suggested criteria, procedures, and a general approach to establish the com-postability of environmentally degradable plastics.

The ASTM has yet to replace this standard.

Biobased – ASTM D6866

The ASTM D6866 method has been developed to certify the biologically derived content of bioplas-tics. Cosmic rays colliding with the atmosphere mean that some of the carbon is the radioactive iso-tope carbon-14. CO_2 from the atmosphere is used by plants in photosynthesis, so new plant material will contain both carbon-14 and carbon-12. Under the right conditions, and over geological times-cales, the remains of living organisms can be transformed into fossil fuels. After ~100,000 years all the carbon-14 present in the original organic material will have undergone radioactive decay leaving only carbon-12. A product made from biomass will have a relatively high level of carbon-14, while a product made from petrochemicals will have no carbon-14. The percentage of renewable carbon in a material (solid or liquid) can be measured with an accelerator mass spectrometer.

There is an important difference between biodegradability and biobased content. A bioplastic such as high-density polyethylene (HDPE) can be 100% biobased (i.e. contain 100% renewable carbon), yet be non-biodegradable. These bioplastics such as HDPE nonetheless play an important role in greenhouse gas abatement, particularly when they are combusted for energy production. The biobased component of these bioplastics is considered carbon-neutral since their origin is from biomass.

Anaerobic Biodegradability – ASTM D5511-02 and ASTM D5526

The ASTM D5511-12 and ASTM D5526-12 are testing methods that comply with international standards such as the ISO DIS 15985 for the biodegradability of plastic.

Engineered Wood

Very large self-supporting wooden roof. Built for the world fair in the year 2000, Hanover, Germany.

75 Unit Apartment building, made largely of wood, in Mission, British Columbia.

Engineered wood, also called composite wood, man-made wood, or manufactured board, includes a range of derivative wood products which are manufactured by binding or fixing the strands, particles, fibers, or veneers or boards of wood, together with adhesives, or other methods of fixation to form composite materials. These products are engineered to precise design specifications which are tested to meet national or international standards. Engineered wood products are used in a variety of applications, from home construction to commercial buildings to industrial products. The products can be used for joists and beams that replace steel in many building projects.

Typically, engineered wood products are made from the same hardwoods and softwoods used to manufacture lumber. Sawmill scraps and other wood waste can be used for engineered wood composed of wood particles or fibers, but whole logs are usually used for veneers, such as plywood, MDF or particle board. Some engineered wood products, like oriented strand board (OSB), can use trees from the poplar family, a common but non-structural species.

Alternatively, it is also possible to manufacture similar engineered bamboo from bamboo; and similar engineered cellulosic products from other lignin-containing materials such as rye straw, wheat straw, rice straw, hemp stalks, kenaf stalks, or sugar cane residue, in which case they contain no actual wood but rather vegetable fibers.

Flat pack furniture is typically made out of man-made wood due to its low manufacturing costs and its low weight, making it easy to transport.

Types of Products

Engineered wood products in a Home Depot store

Plywood

Plywood, wood structural panel, is sometimes called the original engineered wood product. Plywood is manufactured from sheets of cross-laminated veneer and bonded under heat and pressure with durable, moisture-resistant adhesives. By alternating the grain direction of the veneers from layer to layer, or "cross-orienting", panel strength and stiffness in both directions are maximized. Other structural wood panels include oriented strand board and structural composite panels.

Fibreboard

Medium-density fibreboard, is made by breaking down hardwood or softwood residuals into wood fibres, combining it with wax and a resin binder, and forming panels by applying high temperature and pressure.

Oriented Strand Board

Oriented strand board (OSB) is a wood structural panel manufactured from rectangular-shaped strands of wood that are oriented lengthwise and then arranged in layers, laid up into mats, and bonded together with moisture-resistant, heat-cured adhesives. The individual layers are cross-oriented to provide strength and stiffness to the panel. Produced in huge, continuous mats, OSB is a solid panel product of consistent quality with no laps, gaps or voids.

Laminated Timber

Glued laminated timber (glulam) is composed of several layers of dimensional timber glued together with moisture-resistant adhesives, creating a large, strong, structural member that can be used as vertical columns or horizontal beams. Glulam can also be produced in curved shapes, offering extensive design flexibility.

Laminated Veneer

Laminated veneer lumber (LVL) is produced by bonding thin wood veneers together in a large billet. The grain of all veneers in the LVL billet is parallel to the long direction. The resulting product features enhanced mechanical properties and dimensional stability that offer a broader range in product width, depth and length than conventional lumber. LVL is a member of the structural composite lumber (SCL) family of engineered wood products that are commonly used in the same structural applications as conventional sawn lumber and timber, including rafters, headers, beams, joists, rim boards, studs and columns.

Cross Laminated

Cross-Laminated Timber (CLT) is a versatile multi-layered panel made of lumber. Each layer of boards is placed cross-wise to adjacent layers for increased rigidity and strength. CLT can be used for long spans and all assemblies, e.g. floors, walls or roofs. CLT has the advantage of faster construction times as the panels are manufactured and finished off site and supplied ready to fit and screw together as a flat pack assembly project.

Parallel Strand

Parallel strand lumber (PSL) consists of long veneer strands laid in parallel formation and bonded together with an adhesive to form the finished structural section. A strong, consistent material, it has a high load carrying ability and is resistant to seasoning stresses so it is well suited for use as beams and columns for post and beam construction, and for beams, headers, and lintels for light framing construction. PSL is a member of the structural composite lumber (SCL) family of engineered wood products.

Laminated Strand

Laminated strand lumber (LSL) and oriented strand lumber (OSL) are manufactured from flaked wood strands that have a high length-to-thickness ratio. Combined with an adhesive, the strands are oriented and formed into a large mat or billet and pressed. LSL and OSL offer good fastener-holding strength and mechanical connector performance and are commonly used in a variety of applications, such as beams, headers, studs, rim boards, and millwork components. These products are members of the structural composite lumber (SCL) family of engineered wood products. LSL is manufactured from relatively short strands—typically about 1 foot long—compared to the 2 foot to 8 foot long strands used in PSL.

Finger Joint

Finger-jointed lumber is made up of short pieces of wood combined to form longer lengths and is used in doorjambs, mouldings and studs. It is also produced in long lengths and wide dimensions for floors.

Beams

I-joists and wood I-beams are "I"-shaped structural members designed for use in floor and roof construction. An I-joist consists of top and bottom flanges of various widths united with webs of various depths. The flanges resist common bending stresses, and the web provides shear performance. I-joists are designed to carry heavy loads over long distances while using less lumber than a dimensional solid wood joist of a size necessary to do the same task . As of 2005, approximately half of all wood light framed floors were framed using I-joists .

Trusses

Roof trusses and floor trusses are structural frames relying on a triangular arrangement of webs and chords to transfer loads to reaction points. For a given load, long wood trusses built from smaller pieces of lumber require less raw material and make it easier for AC contractors, plumbers, and electricians to do their work, compared to the long 2x10s and 2x12s traditionally used as rafters and floor joists.

Advantages

Engineered wood products are used in a variety of ways, often in applications similar to solid wood products. Engineered wood products may be preferred over solid wood in some applications due to certain comparative advantages:

- Engineered wood is felt to offer structural advantages for home construction.

- Because engineered wood is man-made, it can be designed to meet application-specific performance requirements.

- Engineered wood products are versatile and available in a wide variety of thicknesses, sizes, grades, and exposure durability classifications, making the products ideal for use in unlimited construction, industrial and home project application.

- Engineered wood products are designed and manufactured to maximize the natural strength and stiffness characteristics of wood. The products are very stable and some offer greater structural strength than typical wood building materials.

- Glued laminated timber (glulam) has greater strength and stiffness than comparable dimensional lumber and, pound for pound, is stronger than steel.

- Some engineered wood products offer more design options without sacrificing structural requirements.

- Engineered wood panels are easy to work with using ordinary tools and basic skills. They can be cut, drilled, routed, jointed, glued, and fastened. Plywood can be bent to form curved surfaces without loss of strength. And large panel size speeds construction by reducing the number of pieces to be handled and installed.

- Engineered wood products make more efficient use of wood. They can be made from small pieces of wood, wood that has defects or underutilized species.

- Wooden trusses are competitive in many roof and floor applications, and their high strength-to-weight ratios permit long spans offering flexibility in floor layouts.

- Sustainable design advocates recommend using engineered wood, which can be produced from relatively small trees, rather than large pieces of solid dimensional lumber, which requires cutting a large tree.

Disadvantages

- Some products may burn more quickly than solid lumber.

- They require more primary energy for their manufacture than solid lumber.

- The adhesives used in some products may be toxic. A concern with some resins is the release of formaldehyde in the finished product, often seen with urea-formaldehyde bonded products.

- Cutting and otherwise working with some products can expose workers to toxic compounds.

- Some engineered wood products, such as those specified for interior use, may be weaker and more prone to humidity-induced warping than equivalent solid woods. Most particle and fiber-based boards are not appropriate for outdoor use because they readily soak up water.

Properties

Plywood and OSB typically have a density of 35 to 40 pounds per cubic foot (550 to 650 kg per cubic meter). For example, 3/8" plywood sheathing or OSB sheathing typically has a weight of 1.0 to 1.2 pounds per square foot.

Engineered Wood Flooring Manufacturing

Lamella

The lamella is the face layer of the wood that is visible when installed. Typically, it is a sawn piece of timber. The timber can be cut in three different styles: flat-sawn, quarter-sawn, and rift-sawn. Keep in mind that each cut will give the board a different final appearance.

Core/Substrate

1. Wood ply construction ("sandwich core"): Uses multiple thin plies of wood adhered together. The wood grain of each ply runs perpendicular to the ply below it. Stability is attained from using thin layers of wood that have little to no reaction to climatic change. The wood is further stabilized due to equal pressure being exerted lengthwise and widthwise from the plies running perpendicular to each other.

2. Finger core construction: Finger core engineered wood floors are made of small pieces of milled timber that run perpendicular to the top layer (lamella) of wood. They can be 2-ply or 3-ply, depending on their intended use. If it is three ply, the third ply is often plywood that runs parallel to the lamella. Stability is gained through the grains running perpendicular to each other, and the expansion and contraction of wood is reduced and relegated to the middle ply, stopping the floor from gapping or cupping.

3. Fibreboard: The core is made up of medium or high density fibreboard. Floors with a fibreboard core are hygroscpoic and must never be exposed to large amounts of water or very high humidity - the expansion caused from absorbing water combined with the density of the fibreboard, will cause it to lose its form. Fibreboard is less expensive than timber and can emit higher levels of harmful gases due to its relatively high adhesive content.

4. An engineered flooring construction which is popular in parts of Europe is the hardwood lamella, softwood core laid perpendicular to the lamella, and a final backing layer of the same noble wood used for the lamella. Other noble hardwoods are sometimes used for the back layer but must be compatible. This is thought by many to be the most stable of engineered floors.

Aesthetics

Engineered wood flooring is mainly industrially fabricated in the form of straight edged boards, with milled jointing profiles to provide for interconnecting of the boards. Such manufacturing is most cost efficient but leaves an industrial looking surface. In nature no straight lines exist; therefore there is a rising trend to modify the visual appearance to imitate it. In recent years numerous producers have been taking on the challenge of adding more natural aesthetics.

Adhesives

The types of adhesives used in engineered wood include:

Urea-formaldehyde resins (UF)

> most common, cheapest, and not waterproof.

Phenol formaldehyde resins (PF)

> yellow/brown, and commonly used for exterior exposure products.

Melamine-formaldehyde resins (MF)

> white, heat and water resistant, and often used in exposed surfaces in more costly designs.

Methylene diphenyl diisocyanate (MDI) or polyurethane (PU) resins

> expensive, generally waterproof, and does not contain formaldehyde.

A more inclusive term is *structural composites*. For example, fiber cement siding is made of cement and wood fiber, while cement board is a low density cement panel, often with added resin, faced with fiberglass mesh.

Other Fixations

Some engineered products such as CLT Cross Laminated Timber can be assembled without the use of adhesives using mechanical fixing. These can range from profiled interlocking jointed boards, proprietary metal fixings, nails or timber dowels (Brettstapel - single layer or CLT).

Standards

The following standards are related to engineered wood products:

- EN 300 - Oriented Strand Boards (OSB) — Definitions, classification and specifications
- EN 309 - Particleboards — Definition and classification
- EN 338 - Structural timber - Strength classes
- EN 386 - Glued laminated timber — performance requirements and minimum production requirements
- EN 313-1 - Plywood — Classification and terminology Part 1: Classification
- EN 313-2 - Plywood — Classification and terminology Part 2: Terminology
- EN 314-1 - Plywood — Bonding quality — Part 1: Test methods
- EN 314-2 - Plywood — Bonding quality — Part 2: Requirements
- EN 315 - Plywood — Tolerances for dimensions
- EN 387 - Glued laminated timber — large finger joints - performance requirements and minimum production requirements

- EN 390 - Glued laminated timber — sizes - permissible deviations

- EN 391 - Glued laminated timber — shear test of glue lines

- EN 392 - Glued laminated timber — Shear test of glue lines

- EN 408 - Timber structures — Structural timber and glued laminated timber — Determination of some physical and mechanical properties

- EN 622-1 - Fibreboards — Specifications — Part 1: General requirements

- EN 622-2 - Fibreboards — Specifications — Part 2: Requirements for hardboards

- EN 622-3 - Fibreboards — Specifications — Part 3: Requirements for medium boards

- EN 622-4 - Fibreboards — Specifications — Part 4: Requirements for softboards

- EN 622-5 - Fibreboards — Specifications — Part 5: Requirements for dry process boards (MDF)

- EN 1193 - Timber structures — Structural timber and glued laminated timber - Determination of shear strength and mechanical properties perpendicular to the grain

- EN 1194 - Timber structures — Glued laminated timber - Strength classes and determination of characteristic values

- EN 1995-1-1 - Eurocode 5: Design of timber structures — Part 1-1: General — Common rules and rules for buildings

- EN 12369-1 - Wood-based panels — Characteristic values for structural design — Part 1: OSB, particleboards and fibreboards

- EN 12369-2 - Wood-based panels — Characteristic values for structural design — Part 2: Plywood

- EN 12369-3 - Wood-based panels — Characteristic values for structural design — Part 3: Solid wood panels

- EN 14080 - Timber structures — Glued laminated timber — Requirements

- EN 14081-1 - Timber structures - Strength graded structural timber with rectangular cross section - Part 1: General requirements

Zein

Zein is a class of prolamine protein found in maize (corn). It is usually manufactured as a powder from corn gluten meal. Zein is one of the best understood plant proteins. Pure zein is clear, odorless, tasteless, hard, water-insoluble, and edible, and it has a variety of industrial and food uses.

Ommercial Uses

Historically, zein has been used in the manufacture of a wide variety of commercial products, including coatings for paper cups, soda bottle cap linings, clothing fabric, buttons, adhesives, coatings and binders. The dominant historical use of zein was in the textile fibers market where it was produced under the name "Vicara". With the development of synthetic alternatives, the use of zein in this market eventually disappeared. By using electrospinning, zein fibers have again been produced in the lab, where additional research will be performed to re-enter the fiber market.

Zein's properties make it valuable in processed foods and pharmaceuticals, in competition with insect shellac. It is now used as a coating for candy, nuts, fruit, pills, and other encapsulated foods and drugs. In the United States, it may be labeled as "confectioner's glaze" (which may also refer to shellac-based glazes) and used as a coating on bakery products or as "vegetable protein." It is classified as Generally Recognized as Safe (GRAS) by the U.S. Food and Drug Administration. For pharmaceutical coating, zein is preferred over food shellac, since it is all natural and requires less testing per the USP monographs.

Zein can be further processed into resins and other bioplastic polymers, which can be extruded or rolled into a variety of plastic products. With increasing environmental concerns about synthetic coatings (such as PFOA) and the current higher prices of hydrocarbon-based petrochemicals, there is increased focus on zein as a raw material for a variety of nontoxic and renewable polymer applications, particularly in the paper industry. Other reasons for a renewed interest in zein include concern about the landfill costs of plastics, and consumer interest in natural substances. There are also a number of potential new food industry applications.

Researchers at the University of Illinois at Urbana-Champaign and at William Wrigley Jr. Company have recently been studying the possibility of using zein to replace some of the gum base in chewing gum. They are also studying medical applications such as using the zein molecule to "carry biocompounds to targeted sites in the human body". There are a number of potential food safety applications that may be possible for zein-based packaging according to several researchers. A military contractor is researching the use of zein to protect MRE food packages. Other packaging/food safety applications that have been researched include frozen foods, ready-to-eat chicken, and cheese and liquid eggs. Food researchers in Japan have noted the ability of the zein molecule to act as a water barrier.

While there are numerous existing and potential uses for zein, the main barrier to greater commercial success has been its historic high cost until recently. Zein pricing is now very competitive with food shellac. Zein may be extracted as a byproduct in the manufacturing process for ethanol or in new off-shore manufacture.

Gene Family

Alpha-prolamins are the major seed storage proteins of species of the grass tribe Andropogonea. They are unusually rich in glutamine, proline, alanine, and leucine residues and their sequences show a series of tandem repeats presumed to be the result of multiple intragenic duplication. In *Zea mays* (Maize), the 22 kDa and 19 kDa zeins are encoded by a large multigene family and are the major seed storage proteins accounting for 70% of the total zein fraction. Structurally the 22 kDa and 19 kDa zeins are composed of nine adjacent, topologically antiparallel helices clustered within a distorted cylinder. The 22 kDa alpha-zeins are encoded by 23 genes; twenty-two of the members are found in a roughly tandem array forming a dense gene cluster. The expressed genes in the cluster are interspersed with nonexpressed genes. Interestingly, some of the expressed genes differ in their transcriptional regulation. Gene amplification appears to be in blocks of genes explaining the rapid and compact expansion of the cluster during the evolution of maize.

Other Biodegradable Polymers

- Cellophane

- Plastarch material

- Poly-3-hydroxybutyrate

- Polycaprolactone

- Polyglycolide

- Polylactic acid

Corn Starch

Corn starch, cornstarch, cornflour or maize starch or maize is the starch derived from the corn (maize) grain or wheat. The starch is obtained from the endosperm of the kernel. Corn starch is a popular food ingredient used in thickening sauces or soups, and is used in making corn syrup and other sugars.

History

BROWN & POLSON'S
35 Years
World wide reputation CORN FLOUR

Can be used in all cases where butter and flour thickening for sauces is recommended, and as it requires no butter is to be preferred on account of its plainness.

Advertisement for a Cornflour manufacture, 1894

Cornstarch was discovered in 1840 by Thomas Kingsford, superintendent of a wheat starch factory in Jersey City, New Jersey. Until 1851, corn starch was used primarily for starching laundry and other industrial uses.(It can be used in slime)

Use

Cornstarch is used as a thickening agent in liquid-based foods (e.g., soup, sauces, gravies, custard), usually by mixing it with a lower-temperature liquid to form a paste or slurry. It is sometimes preferred over flour alone because it forms a translucent mixture, rather than an opaque one. As the starch is heated, the molecular chains unravel, allowing them to collide with other starch chains to form a mesh, thickening the liquid (Starch gelatinization).

It is usually included as an anti-caking agent in powdered sugar (10X or confectioner's sugar). Baby powder often includes cornstarch among its ingredients.

Cornstarch when mixed with a fluid can make a non-Newtonian fluid, e.g. adding water makes Oobleck and adding oil makes an Electrorheological fluid.

A common substitute is arrowroot, which replaces cornstarch on a 1:1 ratio.

Cornstarch added to a batter which coated chicken nuggets increased oil absorption and crispness after the latter stages of frying.

Cornstarch can be used to manufacture bioplastics.

Cornstarch is the preferred anti-stick agent on medical products made from natural latex, including condoms, diaphragms and medical gloves. Prior usage of talc was abandoned as talc was believed to be a carcinogen.

Food producers reduce production costs by adding varying amounts of cornstarch to foods, for example to cheese and yogurt. This is more common in the United States of America where the Congress and the Department of Agriculture subsidize and reduce its cost to food manufacturers.

When roasted in a standard oven it produced dextrin, a chemical compound with uses ranging from adhesive to binder for fireworks.

Cornstarch is used to supply glucose to humans who have glycogen storage disease (GSD). Without this, they would not thrive (i.e. little, if any, weight gain) and thus die. Cornstarch can be used starting at age 6 – 12 months which allows feeds to be spaced and glucose fluctuations to be minimized.

Manufacture

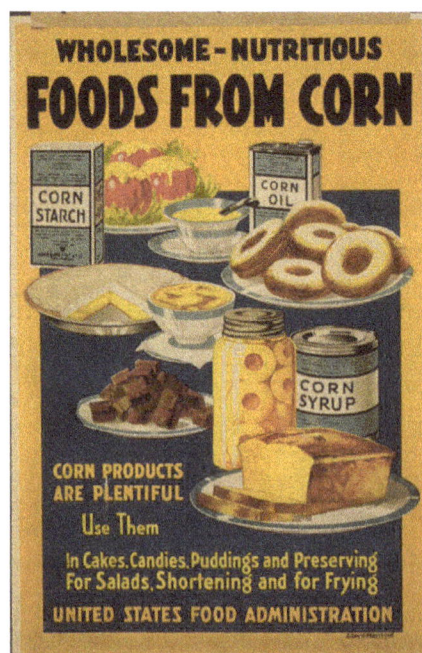

Corn starch shown on a poster, upper left.

The corn is steeped for 30 to 48 hours, which ferments it slightly. The germ is separated from the endosperm and those two components are ground separately (still soaked). Next the starch is removed from each by washing. The starch is separated from the corn steep liquor, the cereal germ, the fibers and the corn gluten mostly in hydrocyclones and centrifuges, and then dried. (The residue from every stage is used in animal feed and to make corn oil or other applications.) This process is called wet milling. Finally, the starch may be modified for specific purposes.

Accident

On June 27, 2015, flammable starch-based powder fueled the Formosa Fun Coast explosion in Taiwan.

Names and Varieties

- Called *cornstarch* in the United States and Canada.

- Called *cornflour* in the United Kingdom, Ireland, Israel and some Commonwealth countries.

- Often called *maizena* in the Netherlands, Belgium, France, Germany, Finland, Austria, Italy, Portugal, Morocco, Brazil, Norway, Denmark, Slovakia, Sweden, Switzerland, Spain, South Africa and Latin America, after the brand.

References

- Starke, Linda (2009). State Of The World 2009: Into a Warming World: a WorldWatch Institute Report on Progress Toward a Sustainable Society. WW Norton & Company. ISBN 978-0-393-33418-0.

- Rafael Auras; Loong-Tak Lim; Susan E. M. Selke; Hideto Tsuji (eds.). Poly(Lactic Acid): Synthesis, Structures, Properties, Processing, and Applications. doi:10.1002/9780470649848. ISBN 9780470293669.

- Manley, Duncan (1998). Biscuit, cookie and cracker manufacturing manuals - Manual 1 - Ingredients. Cambridge, England: Woodhead Publishing Limited. p. 34. ISBN 1 85573 292 0.

- Cho, Renee (2011-08-18). "Is Biomass Really Renewable?". Earth Institute. Columbia University. Retrieved 2016-10-01.

- "NRDC fact sheet lays out biomass basics, campaign calls for action to tell EPA our forests aren't fuel". nrdc.org. Retrieved 28 February 2015.

- REN21 (2011). "Renewables 2011: Global Status Report" (PDF). pp. 13–14. Archived from the original (PDF) on 2011-09-05. Retrieved 2015-01-03.

- "The potential and challenges of drop-in fuels (members only) | IEA Bioenergy Task 39 – Commercializing Liquid Biofuels". task39.sites.olt.ubc.ca. Retrieved 2015-09-10.

- Cotton, Charles A. R.; Jeffrey S. Douglass; Sven De Causmaeker; Katharina Brinkert; Tanai Cardona; Andrea Fantuzzi; A. William Rutherford; James W. Murray (2015).

- "Sweet Sorghum : A New "Smart Biofuel Crop"". Agriculture Business Week. 30 June 2008. Archived from the original on 27 May 2015.

- The National Academies Press (2008). "Water Issues of Biofuel Production Plants". The National Academies Press. Retrieved 18 June 2015.

- Wittbrodt, B., & Pearce, J. M. (2015). The Effects of PLA Color on Material Properties of 3-D Printed Components. Additive Manufacturing. 8, 110–116 (2015). DOI: 10.1016/j.addma.2015.09.006.

7

Understanding Environmental Chemistry

Environmental chemistry is the study of chemical phenomena that occur in the environment. Bio-indicators, aquatic biomonitoring, environmental chemistry and pollinator decline are some of the topics that have been explained in the following chapter.

Environmental Chemistry

Environmental chemistry is the scientific study of the chemical and biochemical phenomena that occur in natural places. It should not be confused with green chemistry, which seeks to reduce potential pollution at its source. It can be defined as the study of the sources, reactions, transport, effects, and fates of chemical species in the air, soil, and water environments; and the effect of human activity and biological activity on these. Environmental chemistry is an interdisciplinary science that includes atmospheric, aquatic and soil chemistry, as well as heavily relying on analytical chemistry and being related to environmental and other areas of science.

White bags filled with contaminated stones line the shore near an industrial oil spill in Raahe, Finland

Environmental chemistry is the study of chemical processes occurring in the environment which are impacted by humankind's activities. These impacts may be felt on a local scale, through the presence of urban air pollutants or toxic substances arising from a chemical waste site, or on a global scale, through depletion of stratospheric ozone or global warming. The focus in our courses and research activities is upon developing a fundamental understanding of the nature of these chemical processes, so that humankind's activities can be accurately evaluated.

Environmental chemistry involves first understanding how the uncontaminated environment works, which chemicals in what concentrations are present naturally, and with what effects. Without this it would be impossible to accurately study the effects humans have on the environment through the release of chemicals.

Environmental chemists draw on a range of concepts from chemistry and various environmental sciences to assist in their study of what is happening to a chemical species in the environment. Important general concepts from chemistry include understanding chemical reactions and equations, solutions, units, sampling, and analytical techniques.

Contamination

A contaminant is a substance present in nature at a level higher than fixed levels or that would not otherwise be there. This may be due to human activity and bioactivety. The term contaminant is often used interchangeably with *pollutant*, which is a substance that has a detrimental impact on the surrounding environment. Whilst a contaminant is sometimes defined as a substance present in the environment as a result of human activity, but without harmful effects, it is sometimes the case that toxic or harmful effects from contamination only become apparent at a later date.

The "medium" (e.g. soil) or organism (e.g. fish) affected by the pollutant or contaminant is called a *receptor*, whilst a *sink* is a chemical medium or species that retains and interacts with the pollutant e.g. as carbon sink and its effects by microbes.

Environmental Indicators

Chemical measures of water quality include dissolved oxygen (DO), chemical oxygen demand (COD), biochemical oxygen demand (BOD), total dissolved solids (TDS), pH, nutrients (nitrates and phosphorus), heavy metals (including copper, zinc, cadmium, lead and mercury), and pesticides.

Applications

Environmental chemistry is used by the Environment Agency (in England and Wales), the United States Environmental Protection Agency, the Association of Public Analysts, and other environmental agencies and research bodies around the world to detect and identify the nature and source of pollutants. These can include:

- Heavy metal contamination of land by industry. These can then be transported into water bodies and be taken up by living organisms.

- Nutrients leaching from agricultural land into water courses, which can lead to algal blooms and eutrophication.

- Urban runoff of pollutants washing off impervious surfaces (roads, parking lots, and rooftops) during rain storms. Typical pollutants include gasoline, motor oil and other hydrocarbon compounds, metals, nutrients and sediment (soil).

- Organometallic compounds.

Methods

Quantitative chemical analysis is a key part of environmental chemistry, since it provides the data that frame most environmental studies.

Common analytical techniques used for quantitative determinations in environmental chemistry include classical wet chemistry, such as gravimetric, titrimetric and electrochemical methods. More sophisticated approaches are used in the determination of trace metals and organic compounds. Metals are commonly measured by atomic spectroscopy and mass spectrometry: Atomic Absorption Spectrophotometry (AAS) and Inductively Coupled Plasma Atomic Emission (ICP-AES) or Inductively Coupled Plasma Mass Spectrometric (ICP-MS) techniques. Organic compounds are commonly measured also using mass spectrometric methods, such as Gas chromatography-mass spectrometry (GC/MS) and Liquid chromatography-mass spectrometry (LC/MS). Tandem Mass spectrometry MS/MS and High Resolution/Accurate Mass spectrometry HR/AM offer sub part per trillion detection. Non-MS methods using GCs and LCs having universal or specific detectors are still staples in the arsenal of available analytical tools.

Other parameters often measured in environmental chemistry are radiochemicals. These are pollutants which emit radioactive materials, such as alpha and beta particles, posing danger to human health and the environment. Particle counters and Scintillation counters are most commonly used for these measurements. Bioassays and immunoassays are utilized for toxicity evaluations of chemical effects on various organisms. Polymerase Chain Reaction PCR is able to identify species of bacteria and other organisms through specific DNA and RNA gene isolation and amplification and is showing promise as a valuable technique for identifying environmental microbial contamination.

Published Analytical Methods

Peer-reviewed test methods have been published by government agencies and private research organizations. Approved published methods must be used when testing to demonstrate compliance with regulatory requirements.

Bioindicator

Caddisfly (order Trichoptera), a macroinvertebrate used as an indicator of water quality.

A bioindicator is any biological species (an "indicator species") or group of species whose function, population, or status can reveal the qualitative status of the environment. For example, copepods and other small water crustaceans that are present in many water bodies can be monitored for changes (biochemical, physiological, or behavioural) that may indicate a problem within their ecosystem. Bioindicators can tell us about the cumulative effects of different pollutants in the ecosystem and about how long a problem may have been present, which physical and chemical testing cannot.

A biological monitor, or biomonitor, can be defined as an organism that provides quantitative information on the quality of the environment around it. Therefore, a good biomonitor will indicate the presence of the pollutant and also attempt to provide additional information about the amount and intensity of the exposure.

Overview

A bio indicator is an organism or biological response that reveals the presence of the pollutants by the occurrence of typical symptoms or measurable responses, and is therefore more qualitative. These organisms (or communities of organisms) deliver information on alterations in the environment or the quantity of environmental pollutants by changing in one of the following ways: physiologically, chemically or behaviourally. The information can be deduced through the study of:

1. their content of certain elements or compounds

2. their morphological or cellular structure

3. metabolic-biochemical processes

4. behaviour, or

5. population structure(s).

The importance and relevance of biomonitors, rather than man-made equipment, is justified by the statement: "There is no better indicator of the status of a species or a system than a species or system itself."

The use of a biomonitor is described as biological monitoring (*abbr.* biomonitoring) and is the use of the properties of an organism to obtain information on certain aspects of the biosphere. Biomonitoring of air pollutants can be passive or active. Passive methods observe plants growing naturally within the area of interest. Active methods detect the presence of air pollutants by placing test plants of known response and genotype into the study area.

Bioaccumulative indicators are frequently regarded as biomonitors.

Depending on the organism selected and their use, there are several types of bio-indicators.

Plant Indicators

The presence or absence of certain plant or other vegetative life in an ecosystem can provide important clues about the health of the environment: environmental preservation.

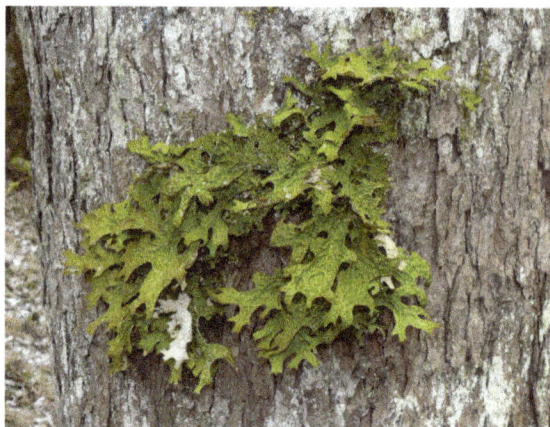
The lichen *Lobaria pulmonaria* is sensitive to air pollution.

There are several types of plant and fungi biomonitors, including mosses, lichens, tree bark, bark pockets, tree rings, leaves, and fungi.

- Lichens are organisms comprising both fungi and algae. They are found on rocks and tree trunks, and they respond to environmental changes in forests, including changes in forest structure – conservation biology, air quality, and climate. The disappearance of lichens in a forest may indicate environmental stresses, such as high levels of sulfur dioxide, sulfur-based pollutants, and nitrogen oxides.

- The composition and total biomass of algal species in aquatic systems serves as an important metric for organic water pollution and nutrient loading such as nitrogen and phosphorus.

There are genetically engineered organisms, that can respond to toxicity levels in the environment; *e.g.*, a type of genetically engineered grass that grows a different colour if there are toxins in the soil.

Animal Indicators and Toxins

An increase or decrease in an animal population may indicate damage to the ecosystem caused by pollution. For example, if pollution causes the depletion of important food sources, animal species dependent upon these food sources will also be reduced in number: population decline. Overpopulation can be the result of opportunistic species growth. In addition to monitoring the size and number of certain species, other mechanisms of animal indication include monitoring the concentration of toxins in animal tissues, or monitoring the rate at which deformities arise in animal populations, or their behaviour either directly in the field or in a lab.

Microbial Indicators and Chemical Pollutants

Microorganisms can be used as indicators of aquatic or terrestrial ecosystem health. Found in large quantities, microorganisms are easier to sample than other organisms. Some microorganisms will produce new proteins, called stress proteins, when exposed to contaminants such as cadmium and benzene. These stress proteins can be used as an early warning system to detect changes in levels of pollution.

Microbial Indicators in Oil and Gas Exploration

Microbial Prospecting for oil and gas (MPOG) is often used to identify prospective areas for oil and gas occurrences. In many cases oil and gas is known to seep toward the surface as a hydrocarbon reservoir will usually leak or have leaked towards the surface through buoyancy forces overcoming sealing pressures. These hydrocarbons can alter the chemical and microbial occurrences found in the near surface soils or can be picked up directly. Techniques used for MPOG include DNA analysis, simple bug counts after culturing a soil sample in a hydrocarbon based medium or by looking at the consumption of hydrocarbon gases in a culture cell.

Microalgae as Bio-indicators for Water Quality

Microalgae have gained attention in the recent years due to several reasons because of their greater sensitivity to pollutants than many other organisms. In addition they occur abundantly in nature, they are an essential component in very many food webs, they are easy to culture and to use in assays and there are few if any ethical issues involved in their use.

Euglena gracilis is a motile freshwater photosynthetic flagellate. Although *Euglena* is rather tolerant to acidity, it responds rapidly and sensitively to environmental stresses such as heavy metals or inorganic and organic compounds. Typical responses are the inhibition of movement and the change of orientation parameters. Moreover, this organism is very easy to handle and grows, making it a very useful tool for eco-toxicological assessments. One very useful particularity of this organism is the gravitactic orientation, which is very sensitive to pollutants.

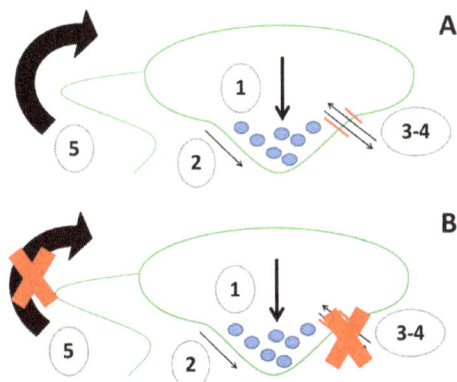

Gravitactic mechanism of the microalgae Euglena gracilis (A) in the absence and (B) in the presence of pollutants.

The gravireceptors are impaired by pollutants such as heavy metals and organic or inorganic compounds. Therefore, the presence of such substances is associated with random movement of the cells in the water column. For short term tests, gravitactic orientation of *E. gracilis* is very sensitive. Other species such as *Paramecium biaurelia* also use gravitactic orientation.

ECOTOX

ECOTOX is an automatic bioassay device used to test the quality of water samples, by the detection of toxic chemicals. It is small piece of hardware containing a miniaturized microscope linked to a camera, an observation cuvette, pumps to mix the water samples with the microalgae; everything being connected to a computer equipped with software. One of the biggest advantages of this device is the automated measurements and analysis, which reduces the risks of personal error. Moreover, it is easy to use, quite cheap and fast: only 10 min are necessary to test a water sample and the corresponding control. Examples of use are the test of seepage water or the determination of the efficiency of purification systems by testing treated waste water before and after purification. The determination of the samples quality is derived from analysis of several parameters related to the movement of the microalgae. All measurements are made automatically with real time image analysis. First the orientation behaviour of the cells is determined using two parameters: the percentage of cells moving upwards giving the direction of the movement and the r-value indicating the precision of the gravitactic orientation which varies from a random movement (r-value=0) to a single direction (r-value=1). Other important parameters are the velocity, the cell motility which represents the percentage of cells moving faster than the minimum velocity and the cell compactness giving information about the shape of the cell. All parameters are compared with a control sample of unpolluted tap water and the percentage of inhibition is calculated. An inhibition indicates the presence of a pollutant. Depending on the aim of the study, the EC50 (the concentration of sample which affects 50 percent of organisms) and the G-value (lowest dilution factor at which no-significant toxic affect can be measured), are calculated. From all those parameters, the gravitactic orientation represented with upward swimming and r-value is the most sensitive.

Macroinvertebrate Bio-indicators

Macroinvertebrates are useful and convenient indicators of the ecological health of a water body. They are almost always present, and are easy to sample and identify. The sensitivity of the range of macroinvertebrates found will enable an objective judgement of the ecological condition to be made. Tolerance values are commonly used to assess water pollution.

In Australia, the SIGNAL method has been developed and is used by researchers and community "Waterwatch" groups to monitor water health.

In Europe, a remote online biomonitoring system was designed in 2006. It is based on bivalve molluscs and the exchange of real time data between a remote intelligent device in the field (able to work for more than 1 year without *in-situ* human intervention) and a data centre designed to capture, process and distribute on the web information derived from the data. The technique relates bivalve behaviour, specifically shell gaping activity, to water quality changes. This technology has been successfully used for the assessment of coastal water quality in various countries (France, Spain, Norway, Russia, Svalbard (Ny Alesund) and New Caledonia).

In the United States, the Environmental Protection Agency (EPA) has published *Rapid Bioassessment Protocols,* based on macroinvertebrates, as well as periphyton and fish. These protocols are used by many federal, state and local government agencies to design biosurveys for assessment of water quality. Volunteer stream monitoring organizations around the U.S., working in cooperation with government agencies, typically use macroinvertebrate methods. The species identification procedures are conducted in the field without the use of specialized equipment, and the techniques can be easily taught in volunteer training sessions.

In South Africa, the Southern African Scoring System (SASS) method was developed as a rapid bioassessment technique, based on benthic macroinvertebrates, and is used for the assessment of water quality in Southern African rivers. The SASS aquatic biomonitoring tool has been refined over the past 30 years and is now on the fifth version (SASS5) which has been specifically modified in accordance with international standards, namely the ISO/IEC 17025 protocol. The SASS5 method is used by the South African Department of Water Affairs as a standard method for River Health Assessment, which feeds the national River Health Programme and the national Rivers Database.

The imposex phenomenon in the dog conch species of sea snail leads to the abnormal development of a penis in females, but does not cause sterility. Because of this, the species has been suggested as a good indicator of pollution with organic man-made tin compounds in Malaysian ports.

Aquatic Biomonitoring

Aquatic biomonitoring is the science of inferring the ecological condition of rivers, lakes, streams, and wetlands by examining the organisms that live there. While aquatic biomonitoring is the most common form of such biomonitoring, any ecosystem can be studied in this manner.

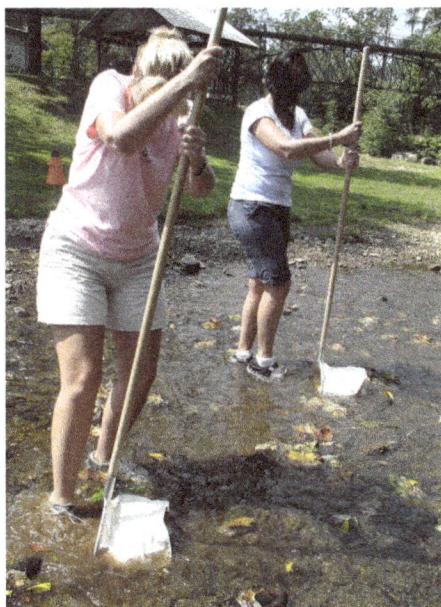

A biosurvey on the North Toe River. North Carolina

Biomonitoring typically takes different approaches:

- *Bioassays*, where test organisms are exposed to an environment to see if mutations or deaths occur. Typical organisms used in bioassays are fish, water fleas (Daphnia), and frogs.

- *Community assessments*, also called *biosurveys,* where an entire community of organisms is sampled, to see what types of taxa remain. In aquatic ecosystems, these assessments often focus on invertebrates, algae, macrophytes (aquatic plants), fish, or amphibians. Rarely, other large vertebrates (reptiles, birds, and mammals) are considered as well.

- *Online biomonitoring devices*, using the ability of animals to permanently taste their environment. Different types of animals are used for that purpose either under lab or field conditions. The use of valve opening/closing activity of clams is one of the possible ways to monitor *in-situ* the quality of freshwater and coastal waters.

Aquatic invertebrates have the longest history of use in biomonitoring programs. In typical unpolluted temperate streams of Europe and North America, certain insect taxa predominate. Mayflies (Ephemeroptera), caddisflies (Trichoptera), and stoneflies (Plecoptera) are the most common insects in these undisturbed streams. In rivers disturbed by urbanization, agriculture, forestry, and other perturbations, flies (Diptera), and especially midges (family Chironomidae) predominate. Aquatic invertebrates are responsive to climate change.

Environmental Chemistry

Environmental chemistry is the scientific study of the chemical and biochemical phenomena that occur in natural places. It should not be confused with green chemistry, which seeks to reduce potential pollution at its source. It can be defined as the study of the sources, reactions, transport, effects, and fates of chemical species in the air, soil, and water environments; and the effect of human activity and biological activity on these. Environmental chemistry is an interdisciplinary science that includes atmospheric, aquatic and soil chemistry, as well as heavily relying on analytical chemistry and being related to environmental and other areas of science.

Environmental chemistry is the study of chemical processes occurring in the environment which are impacted by humankind's activities. These impacts may be felt on a local scale, through the presence of urban air pollutants or toxic substances arising from a chemical waste site, or on a global scale, through depletion of stratospheric ozone or global warming. The focus in our courses and research activities is upon developing a fundamental understanding of the nature of these chemical processes, so that humankind's activities can be accurately evaluated.

Environmental chemistry involves first understanding how the uncontaminated environment works, which chemicals in what concentrations are present naturally, and with what effects. Without this it would be impossible to accurately study the effects humans have on the environment through the release of chemicals.

Environmental chemists draw on a range of concepts from chemistry and various environmental sciences to assist in their study of what is happening to a chemical species in the environment. Important general concepts from chemistry include understanding chemical reactions and equations, solutions, units, sampling, and analytical techniques.

Contamination

A contaminant is a substance present in nature at a level higher than fixed levels or that would not otherwise be there. This may be due to human activity and bioactivety. The term contaminant is often used interchangeably with *pollutant*, which is a substance that has a detrimental impact on the surrounding environment. Whilst a contaminant is sometimes defined as a substance present in the environment as a result of human activity, but without harmful effects, it is sometimes the case that toxic or harmful effects from contamination only become apparent at a later date.

The "medium" (e.g. soil) or organism (e.g. fish) affected by the pollutant or contaminant is called a *receptor*, whilst a *sink* is a chemical medium or species that retains and interacts with the pollutant e.g. as carbon sink and its effects by microbes.

Environmental Indicators

Chemical measures of water quality include dissolved oxygen (DO), chemical oxygen demand (COD), biochemical oxygen demand (BOD), total dissolved solids (TDS), pH, nutrients (nitrates and phosphorus), heavy metals (including copper, zinc, cadmium, lead and mercury), and pesticides.

Applications

Environmental chemistry is used by the Environment Agency (in England and Wales), the United States Environmental Protection Agency, the Association of Public Analysts, and other environmental agencies and research bodies around the world to detect and identify the nature and source of pollutants. These can include:

- Heavy metal contamination of land by industry. These can then be transported into water bodies and be taken up by living organisms.

- Nutrients leaching from agricultural land into water courses, which can lead to algal blooms and eutrophication.

- Urban runoff of pollutants washing off impervious surfaces (roads, parking lots, and rooftops) during rain storms. Typical pollutants include gasoline, motor oil and other hydrocarbon compounds, metals, nutrients and sediment (soil).

- Organometallic compounds.

Methods

Quantitative chemical analysis is a key part of environmental chemistry, since it provides the data that frame most environmental studies.

Common analytical techniques used for quantitative determinations in environmental chemistry include classical wet chemistry, such as gravimetric, titrimetric and electrochemical methods. More sophisticated approaches are used in the determination of trace metals and organic compounds. Metals are commonly measured by atomic spectroscopy and mass spectrometry: Atomic Absorption Spectrophotometry (AAS) and Inductively Coupled Plasma Atomic Emission (ICP-AES) or Inductively Coupled Plasma Mass Spectrometric (ICP-MS) techniques. Organic compounds are commonly measured also using mass spectrometric methods, such as Gas chromatography-mass spectrometry (GC/MS) and Liquid chromatography-mass spectrometry (LC/MS). Tandem Mass spectrometry MS/MS and High Resolution/Accurate Mass spectrometry HR/AM offer sub part per trillion detection. Non-MS methods using GCs and LCs having universal or specific detectors are still staples in the arsenal of available analytical tools.

Other parameters often measured in environmental chemistry are radiochemicals. These are pollutants which emit radioactive materials, such as alpha and beta particles, posing danger to human health and the environment. Particle counters and Scintillation counters are most commonly used for these measurements. Bioassays and immunoassays are utilized for toxicity evaluations of chemical effects on various organisms. Polymerase Chain Reaction PCR is able to identify species of bacteria and other organisms through specific DNA and RNA gene isolation and amplification and is showing promise as a valuable technique for identifying environmental microbial contamination.

Published Analytical Methods

Peer-reviewed test methods have been published by government agencies and private research organizations. Approved published methods must be used when testing to demonstrate compliance with regulatory requirements.

Endophenotype

Endophenotype is a genetic epidemiology term which is used to separate behavioral symptoms into more stable phenotypes with a clear genetic connection. The concept was coined by Bernard John and Kenneth R. Lewis in a 1966 paper attempting to explain the geographic distribution of grasshoppers. They claimed that the particular geographic distribution could not be explained by the obvious and external "exophenotype" of the grasshoppers, but instead must be explained by their microscopic and internal "endophenotype."

The next major use of the term was in psychiatric genetics, to bridge the gap between high-level symptom presentation and low-level genetic variability, such as single nucleotide polymorphisms. It is therefore more applicable to more heritable disorders, such as bipolar disorder and schizophrenia. Since then, the concept has expanded to many other fields, such as the study of ADHD, addiction, Alzheimer's disease obesity and cystic fibrosis. Some other terms which have a similar meaning but do not stress the genetic connection as highly are "intermediate phenotype", "biological marker", "subclinical trait", "vulnerability marker", and "cognitive marker". The strength of an endophenotype is its ability to differentiate between potential diagnoses that present with similar symptoms.

Definition

In psychiatry research, the accepted criteria which a biomarker must fulfill to be called an endophenotype include:

1. An endophenotype must segregate with illness in the population.

2. An endophenotype must be heritable.

3. An endophenotype must not be state-dependent (i.e., manifests whether illness is active or in remission).

4. An endophenotype must co-segregate with illness within families.

5. An endophenotype must be present at a higher rate within affected families than in the population.

6. An endophenotype must be amenable to reliable measurement, and be specific to the ill-ness of interest.

For Schizophrenia

In the case of schizophrenia, the overt symptom could be a psychosis, but the underlying phenotypes are, for example, a lack of sensory gating and a decline in working memory. Both of these traits have a clear genetic component and can thus be called endophenotypes. A strong candidate for schizophrenia endophenotype is prepulse inhibition, the ability to inhibit the reaction to startling stimuli.

Some distinct genes that could underlie certain endophenotypic traits in schizophrenia include:

- RELN – coding the reelin protein downregulated in patients' brains. In one 2008 study its variants were associated with performance in verbal and visual working memory tests in the nuclear families of the sufferers.

- FABP7, coding the *Fatty acid-binding protein 7 (brain)*, one SNP of which was associated with schizophrenia in one 2008 study, is also linked to prepulse inhibition in mice. It is still uncertain though whether the finding will be replicated for human patients.

- CHRNA7, coding the neuronal nicotinic acetylcholine receptor alpha7 subunit. alpha7-containing receptors are known to improve prepulse inhibition, pre-attentive and attentive states.

For Bipolar Disorder

In bipolar disorder, one commonly identified endophenotype is a deficit in face emotion labeling, which is found in both individuals with bipolar disorder and in individuals who are "at risk" (i.e., have a first degree relative with bipolar disorder). Using fMRI, this endophenotype has been linked to dysfunction in the dorsolateral and ventrolateral prefrontal cortex, anterior cingulate cortex, striatum, and amygdala. A polymorphism in the *CACNA1C* gene coding for the voltage-dependent calcium channel $Ca_v1.2$ has been found to be associated with deficits in facial emotion recognition.

For Suicide

The endophenotype concept has also been used in suicide studies. Personality characteristics can be viewed as endophenotypes that may exert a diathesis effect on an individual's susceptibility to suicidal behavior. Although the exact identification of these endophenotypes is controversial, certain traits such as impulsivity and aggression are commonly cited risk factors. One such genetic basis for one of these at-risk endophenotypes has been suggested in 2007 to be the gene coding for the serotonin receptor 5-HT_{1B}, known to be relevant in aggressive behaviors.

Pollinator Decline

The term pollinator decline refers to the reduction in abundance of insect and other animal pollinators in many ecosystems worldwide beginning at the end of the twentieth century, and continuing into the present day.

Pollinators participate in sexual reproduction of many plants, by ensuring cross-pollination, essential for some species, or a major factor in ensuring genetic diversity for others. Since plants are the primary food source for animals, the reduction of one of the primary pollination agents, or even their possible disappearance, has raised concern, and the conservation of pollinators has become part of biodiversity conservation efforts.

Consequences

The value of bee pollination in human nutrition and food for wildlife is immense and difficult to quantify.

60 to 80% of the world's flowering plant species are animal pollinated, and 35% of crop production and 60% of crop plant species depend on animal pollinators. It is commonly said that about one third of human nutrition is due to bee pollination. This includes the majority of fruits, many vegetables (or their seed crop) and secondary effects from legumes such as alfalfa and clover fed to livestock.

In 2000, Drs. Roger Morse and Nicholas Calderone of Cornell University attempted to quantify the effects of just one pollinator, the Western honey bee, on only US food crops. Their calculations came up with a figure of US $14.6 billion in food crop value. In 2009, another study calculated the worldwide value of pollination to agriculture. They calculated the costs using the proportion of each of 100 crops that need pollinators that would not be produced in case insect pollinators disappeared completely. The economic value of insect pollination was then of €153 billion.

Increasing Public Awareness

There are international initiatives (e.g. the International Pollinator Initiative (IPI)) that highlight the need for public participation and awareness of pollinator, such as bees, conservation

Possible Explanations

Pesticide Misuse

Studies have linked neonicotinoid pesticide exposure to bee health decline. These studies add to a growing body of scientific literature and strengthen the case for removing pesticides toxic to bees from the market. Pesticides interfere with honey bee brains, affecting their ability to navigate. Pesticides prevent bumble bees from collecting enough food to produce new queens.

Neonicotinoids are highly toxic to a range of insects, including honey bees and other pollinators. They are taken up by a plant's vascular system and expressed through pollen, nectar and guttation droplets from which bees forage and drink. They are particularly dangerous because, in addition to being acutely toxic in high doses, they also result in serious sub-lethal effects when insects are exposed to chronic low doses, as they are through pollen and water droplets laced with the chemical as well as dust that is released into the air when coated seeds are planted. These effects cause significant problems for the health of individual honey bees as well as the overall health of honey bee colonies and they include disruptions in mobility, navigation, feeding behavior, foraging activity, memory and learning, and overall hive activity.

A French 2012 study of *Apis mellifera* (western honey bee or European honey bee) that focused on the neonicotinoid pesticide thiamethoxam, which is metabolized by bees into clothianidin, a pesticide cited in legal action, tested the hypothesis that a sub-lethal exposure to a neonicotinoid indirectly increases hive death rate through homing failure in foraging honey bees. When exposed to sub-lethal doses of thiamethoxam, at levels present in the environment, honey bees were less likely to return to the hive after foraging than control bees that were tracked with Radio-Frequency Identification (RFID) tagging technology, but not intentionally dosed with pesticides. Higher risks are observed when the homing task is more challenging. The survival rate is even lower when exposed bees are placed in foraging areas with which they are less familiar.

In their 2014 study of *Bombus terrestris* (buff-tailed bumblebee or large earth bumblebee), researchers tracked bees using RFID tagging technology, and found that a sub-lethal exposure to either imidacloprid (a neonicotinoid) and/or a pyrethroid (?-cyhalothrin) over a four-week period caused impairment of the bumble bee's ability to forage.

Imidacloprid effects on bees were examined by researchers exposing colonies of bumble bees to levels of imidacloprid that are realistic in the natural environment, then allowed them to develop under field conditions. Treated colonies had a significantly reduced growth rate and suffered an 85% reduction in production of new queens compared to unexposed control colonies. The study shows that bumble bees, which are wild pollinators, are suffering similar impacts of pesticide exposure to "managed" honey bees. Wild pollinators provide ecosystem services both in agriculture and to a wide range of wild plants that could not survive without insect pollination.

In March, 2012, commercial beekeepers and environmental organizations filed an emergency legal petition with the U.S. Environmental Protection Agency (EPA) to suspend use of clothianidin, urging the agency to adopt safeguards. The legal petition is supported by over one million citizen petition signatures, targets the pesticide for its harmful impacts on honey bees. The petition points to the fact that the EPA failed to follow its own regulations. EPA granted a conditional, or temporary, registration of clothianidin in 2003 without a field study establishing that the pesticide would have no "unreasonable adverse effects" on pollinators. The conditional registration was contingent upon the submission of an acceptable field study, but this requirement has not been met. EPA continues to allow the use of clothianidin nine years after acknowledging that it had an insufficient legal basis for initially allowing its use. Additionally, the product labels on pesticides containing clothianidin are inadequate to prevent excessive damage to non-target organisms, which is a violation of the requirements for using a pesticide and further warrants removing all such mislabeled pesticides from use.

The disappearance of honeybees was documented in the 2009 film *Vanishing of the Bees* by George Langworthy and Maryam Henein.

Rapid Transfer of Parasites and Diseases of Pollinator Species Around the World

Increased international commerce has moved diseases of the honey bee such as American foulbrood and chalkbrood, and parasites such as varroa mites, acarina mites, and the small African hive beetle to new areas of the world, causing much loss of bees in the areas where they do not have much resistance to these pests. Imported fire ants have decimated ground nesting bees in wide areas of the southern US.

Loss of Habitat and Forage

Bees and other pollinators face a higher risk of extinction due to loss of habitat and access to natural food sources. The global dependency on livestock and agriculture has rendered no less than 50% of the earths landmass uninhabitable for bees. The agricultural practice of planting one crop (monoculture) in a given area year after year leads to extreme malnourishment. Regardless if the planted crop does flower and provide food for the bee, the bee will still be malnourished because a single plant cannot meet the nutrient requirements. Furthermore, the crops needed to support livestock (primarily cattle) tend to be grains which do not provide nectar.

Air Pollution

Researchers at the University of Virginia have discovered that air pollution from automobiles and power plants has been inhibiting the ability of pollinators such as bees and butterflies to find the fragrances of flowers. Pollutants such as ozone, hydroxyl, and nitrate radicals bond quickly with volatile scent molecules of flowers, which consequently travel shorter distances intact. There results a vicious cycle in which pollinators travel increasingly longer distances to find flowers providing them nectar, and flowers receive inadequate pollination to reproduce and diversify.

Changes in Seasonal Behaviour due to Global Warming

In 2014 the Intergovernmental Panel on Climate Change reported that bees, butterflies and other pollinators faced increased risk of extinction because of global warming due to alterations in the seasonal behaviour of species. Climate change was causing bees to emerge at different times in the year when flowering plants were not available.

The structure of Plant-pollinator Networks

Wild pollinators often visit a large number of plant species and plants are visited by a large number of pollinator species. All these relations together form a network of interactions between plants and pollinators. Surprising similarities were found in the structure of networks consisting out of the interactions between plants and pollinators. This structure was found to be similar in very different ecosystems on different continents, consisting of entirely different species.

The structure of plant-pollinator networks may have large consequences for the way in which pollinator communities respond to increasingly harsh conditions. Mathematical models, examining the consequences of this network structure for the stability of pollinator communities suggest that the specific way in which plant-pollinator networks are organized minimizes competition between pollinators and may even lead to strong indirect facilitation between pollinators when conditions are harsh. This makes that pollinator species together can survive under harsh conditions. But it also means that pollinator species collapse simultaneously when conditions pass a critical point. This simultaneous collapse occurs, because pollinator species depend on each other when surviving under difficult conditions.

Such a community-wide collapse, involving many pollinator species, can occur suddenly when increasingly harsh conditions pass a critical point and recovery from such a collapse might not be easy. The improvement in conditions needed for pollinators to recover, could be substantially larger than the improvement needed to return to conditions at which the pollinator community collapsed.

Solutions

Conservation and Restoration

Efforts are being made to sustain pollinator diversity in agro and natural eco-systems by some environmental groups. Prairie restoration, establishment of wildlife preserves, and encouragement of diverse wildlife landscaping rather than mono culture lawns, are examples of ways to help pollinators.

In June 2014 the Obama administration published a fact sheet "The Economic Challenge Posed by Declining Pollinator Populations", which stated that "President's 2015 Budget recommends approximately $50 million across multiple agencies within USDA to ... strengthen pollinator habitat in core areas, double the number of acres in the Conservation Reserve Program that are dedicated to pollinator health ...".

Research

The Obama administration's 2015 Budget also recommended to "enhance research at USDA and through public-private grants, ... and increase funding for surveys to determine the impacts on pollinator losses".

SmartBees is a European research project of 16 entities (universities, research institutions and companies) funded by the EU, headquartered in Berlin. Its goal is to elicit causes of resistance to CCD, develop breeding to increase CCD resistance and to counteract the replacement of many native European bees with only two specific races.

CoLOSS (Prevention of honey bee COlony LOSSes) is an international, non-profit association headquartered in Bern, Switzerland to "improve the well-being of bees at a global level", composed of researchers, veterinarians, agriculture extension specialists, and students from 69 countries. Their 3 core projects are standardization of methods for studying the honey bee, colony loss monitoring and B-RAP (Bridging Research and Practice).

Contract Pollination

The decline of pollinators is compensated to some extent by beekeepers becoming migratory, following the bloom northward in the spring from southern wintering locations. Migration may be for traditional honey crops, but increasingly is for contract pollination to supply the needs for growers of crops that require it.

References

- Harrison, R.M (edited by). Understanding Our Environment, An Introduction to Environmental Chemistry and Pollution, Third Edition. Royal Society of Chemistry. 1999. ISBN 0-85404-584-8

- Sigel, A. (2010). Sigel, H.; Sigel, R.K.O., eds. Organometallics in Environment and Toxicology. Metal Ions in Life Sciences. 7. Cambridge: RSC publishing. ISBN 978-1-84755-177-1.

- Gooderham, John; Tsyrlin, Edward (2002). The Waterbug Book: A Guide to the Freshwater Macroinvertebrates of Temperate Australia. Collingswood, Victoria: CSIRO Publishing. ISBN 0 643 06668 3.

- Harrison, R.M (edited by). Understanding Our Environment, An Introduction to Environmental Chemistry and Pollution, Third Edition. Royal Society of Chemistry. 1999. ISBN 0-85404-584-8

- Sigel, A. (2010). Sigel, H.; Sigel, R.K.O., eds. Organometallics in Environment and Toxicology. Metal Ions in Life Sciences. 7. Cambridge: RSC publishing. ISBN 978-1-84755-177-1.

- NCSU Water Quality Group. "Biomonitoring". WATERSHEDSS: A Decision Support System for Nonpoint Source Pollution Control. Raleigh, NC: North Carolina State University. Retrieved 2016-07-31.

- "MolluScan Eye". Environnements et Paléoenvironnements Océaniques et Continentaux. Talence, France: Université de Bordeaux. Retrieved 2016-08-04.

- The Value of Honey Bees As Pollinators of U.S. Crops in 2000, Drs. Roger Morse and Nicholas Calderone of Cornell University (2000) : "Archived copy" (PDF). Archived from the original (PDF) on 2014-07-22. Retrieved 2016-02-08.

- "Biological Stream Monitoring". Izaak Walton League of America. Archived from the original on 2015-04-21. Retrieved 2010-08-14.

- Office of the Press Secretary (June 20, 2014). "The Economic Challenge Posed by Declining Pollinator Populations" (Factsheet). The White House. Retrieved 31 August 2015.

- Gosden Emily (29 March 2014) Bees and the crops they pollinate are at risk from climate change, IPCC report to warn The Daily Telegraph, Retrieved 30 March 2014

- Gill, Richard J.; Raine, Nigel E. (7 July 2014). "Chronic impairment of bumblebee natural foraging behaviour induced by sublethal pesticide exposure". Functional Ecology. 28 (6): 1459–1471. doi:10.1111/1365-2435.12292.

8

Click Chemistry: An Integrated Study

Click chemistry is also known as tagging; it is not a particular reaction but it helps in describing a way to generate products that follow the examples of nature. Biocompatibility, bioconjugation and azide-alkyne huidgen cycloaddition are explained in the following section. The chapter focuses on all the aspects related to click chemistry and helps the readers in understanding the subject matter.

Click Chemistry

In chemical synthesis, "click" chemistry, more commonly called tagging, is a class of biocompatible reactions intended primarily to join substrates of choice with specific biomolecules. Click chemistry is not a single specific reaction, but describes a way of generating products that follows examples in nature, which also generates substances by joining small modular units. In general, click reactions usually join a biomolecule and a reporter molecule. Click chemistry is not limited to biological conditions: the concept of a "click" reaction has been used in pharmacological and various biomimetic applications. However, they have been made notably useful in the detection, localization and qualification of biomolecules.

Click reactions occur in one pot, are not disturbed by water, generate minimal and inoffensive byproducts, and are "spring-loaded"—characterized by a high thermodynamic driving force that drives it quickly and irreversibly to high yield of a single reaction product, with high reaction specificity (in some cases, with both regio- and stereo-specificity). These qualities make click reactions particularly suitable to the problem of isolating and targeting molecules in complex biological environments. In such environments, products accordingly need to be physiologically stable and any byproducts need to be non-toxic (for *in vivo* systems).

By developing specific and controllable bioorthogonal reactions, scientists have opened up the possibility of hitting particular targets in complex cell lysates. Recently, scientists have adapted click chemistry for use in live cells, for example using small molecule probes that find and attach to their targets by click reactions. Despite challenges of cell permeability, bioorthogonality, background labeling, and reaction efficiency, click reactions have already proven useful in a new generation of pulldown experiments (in which particular targets can be isolated using, for instance, reporter molecules which bind to a certain column), and fluorescence spectrometry (in which the fluorophore is attached to a target of interest and the target quantified or located). More recently, novel methods have been used to incorporate click reaction partners onto and into biomolecules, including the incorporation of unnatural amino acids containing reactive groups into proteins and the modification of nucleotides. These techniques represent a part of the field of chemical biology, in which click chemistry plays a fundamental role by intentionally and specifically coupling modular units to various ends.

The term "click chemistry" was coined by K. Barry Sharpless in 1998, and was first fully described by Sharpless, Hartmuth Kolb, and M.G. Finn of The Scripps Research Institute in 2001.

Background

Click chemistry is a method for attaching a probe or substrate of interest to a specific biomolecule, a process called bioconjugation. The possibility of attaching fluorophores and other reporter molecules has made click chemistry a very powerful tool for identifying, locating, and characterizing both old and new biomolecules.

One of the earliest and most important methods in bioconjugation was to express a reporter on the same open reading frame as a biomolecule of interest. Notably, GFP was first (and still is) expressed in this way at the N- or C- terminus of many proteins. However, this approach comes with several difficulties. For instance, GFP is a very large unit and can often affect the folding of the protein of interest. Moreover, by being expressed at either terminus, the GFP adduct can also affect the targeting and expression of the desired protein. Finally, using this method, GFP can only be attached to proteins, and not post-translationally, leaving other important biomolecular classes (nucleic acids, lipids, carbohydrates, etc.) out of reach.

To overcome these challenges, chemists have opted to proceed by identifying pairs of bioorthogonal reaction partners, thus allowing the use of small exogenous molecules as biomolecular probes. A fluorophore can be attached to one of these probes to give a fluorescence signal upon binding of the reporter molecule to the target—just as GFP fluoresces when it is expressed with the target.

Now limitations emerge from the chemistry of the probe to its target. In order for this technique to be useful in biological systems, click chemistry must run at or near biological conditions, produce little and (ideally) non-toxic byproducts, have (preferably) single and stable products at the same conditions, and proceed quickly to high yield in one pot. Existing reactions, such as Staudinger ligation and the Huisgen 1,3–dipolar cycloaddition, have been modified and optimized for such reaction conditions. Today, research in the field concerns not only understanding and developing new reactions and repurposing and re-understanding known reactions, but also expanding methods used to incorporate reaction partners into living systems, engineering novel reaction partners, and developing applications for bioconjugation.

Reactions

According to Sharpless, a desirable click chemistry reaction would:

- be modular

- be wide in scope

- give very high chemical yields

- generate only inoffensive byproducts

- be stereospecific

- be physiologically stable

- exhibit a large thermodynamic driving force (> 84 kJ/mol) to favor a reaction with a single reaction product. A distinct exothermic reaction makes a reactant "spring-loaded".

- have high atom economy.

The process would preferably:

- have simple reaction conditions

- use readily available starting materials and reagents

- use no solvent or use a solvent that is benign or easily removed (preferably water)

- provide simple product isolation by non-chromatographic methods (crystallisation or distillation)

Many of the click chemistry criteria are subjective, and even if measurable and objective criteria could be agreed upon, it is unlikely that any reaction will be perfect for every situation and application. However, several reactions have been identified that fit the concept better than others:

- [3+2] cycloadditions, such as the Huisgen 1,3-dipolar cycloaddition, in particular the Cu(I)-catalyzed stepwise variant, are often referred to simply as Click reactions

- Thiol-ene reaction

- Diels-Alder reaction and inverse electron demand Diels-Alder reaction

- [4+1] cycloadditions between isonitriles (isocyanides) and tetrazines

- nucleophilic substitution especially to small strained rings like epoxy and aziridine compounds

- carbonyl-chemistry-like formation of ureas but not reactions of the aldol type due to low thermodynamic driving force.

- addition reactions to carbon-carbon double bonds like dihydroxylation or the alkynes in the thiol-yne reaction.

Copper(I)-Catalyzed Azide-Alkyne Cycloaddition (CuAAC)

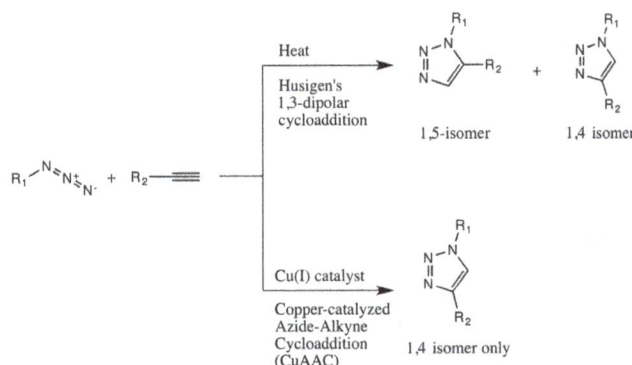

A comparison of the Huisgen and the copper-catalyzed Azide-Alkyne cycloadditions

The classic click reaction is the Copper-catalyzed reaction of an azide with an alkyne to form a 5-membered heteroatom ring: a Cu(I)-Catalyzed Azide-Alkyne Cycloaddition (CuAAC). The first triazole synthesis, from diethyl acetylenedicarboxylate and phenyl azide, was reported by Arthur Michael in 1893. Later, in the middle of the 20th century, this family of 1,3-dipolar cycloadditions took on Huisgen's name after his studies of their reaction kinetics and conditions.

The Copper(I)-catalysis of the Huisgen 1,3-dipolar cycloaddition was discovered concurrently and independently by the groups of Valery V. Fokin and K. Barry Sharpless at the Scripps Research Institute in California and Morten Meldal in the Carlsberg Laboratory, Denmark. The copper-catalyzed version of this reaction gives only the 1,4-isomer, whereas Huisgen's non-catalyzed 1,3-dipolar cycloaddition gives both the 1,4- and 1,5-isomers, is slow, and requires a temperature of 100 degrees Celsius.

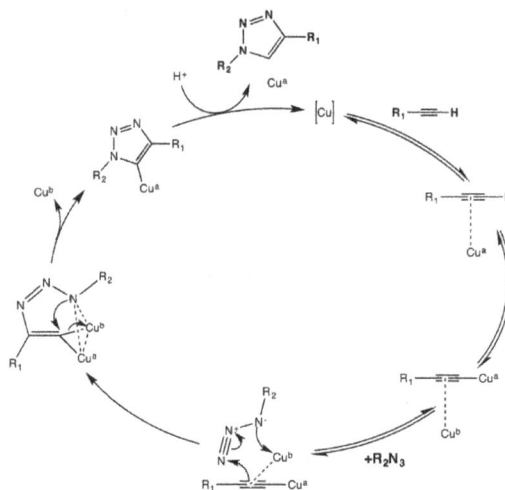

The two-copper mechanism of the CuAAC catalytic cycle

Moreover, this copper-catalyzed "click" does not require ligands on the metal, although accelerating ligands such as Tris(triazolyl)methyl amine ligands with various substituents have been reported and used with success in aqueous solution. Other ligands such as PPh3 and TBIA can also be used, even though PPh3 is liable to Staudinger ligation with the azide substituent. Cu_2O in water at room temperature was found also to catalyze the same reaction in 15 minutes with 91% yield.

The first reaction mechanism proposed included one catalytic copper atom, but isotope studies have suggested the contribution of two functionally distinct Cu atoms in the CuAAC mechanism. Even though this reaction proceeds effectively at biological conditions, copper in this range of dosage is cytotoxic. Solutions to this problem have been presented, such as using water-soluble ligands on the copper to enhance cell penetration of the catalyst and thereby reduce the dosage needed, or to use chelating ligands to further increase the effective concentration of Cu(I) and thereby decreasing the actual dosage.

Although the Cu(I)-catalyzed variant was first reported by Meldal and co-workers for the synthesis of peptidotriazoles on solid support, they needed more time to discover the full scope of the reaction and were overtaken by the publicly more recognized Sharpless. Meldal and co-workers also chose not to label this reaction type "click chemistry" which allegedly caused their discovery to be largely overlooked by the mainstream chemical society. Sharpless and Fokin independently described it as a reliable catalytic process offering "an unprecedented level of selectivity, reliability, and scope for those organic synthesis endeavors which depend on the creation of covalent links between diverse building blocks."

An analogous RuAAC reaction catalyzed by ruthenium, instead of copper, was reported by the Sharpless group in 2005, and allows for the selective production of 1,5-isomers.

Strain-promoted Azide-Alkyne Cycloaddition (SPAAC)

The Bertozzi group further developed one of Huisgen's copper-free click reactions to overcome the cytotoxicity of the CuAAC reaction. Instead of using Cu(I) to activate the alkyne, the alkyne is instead introduced in a strained difluorooctyne (DIFO), in which the electron-withdrawing, propargylic, gem-fluorines act together with the ring strain to greatly destabilize the alkyne. This destabilization increasing the reaction driving force, and the desire of the cycloalkyne to relieve its ring strain.

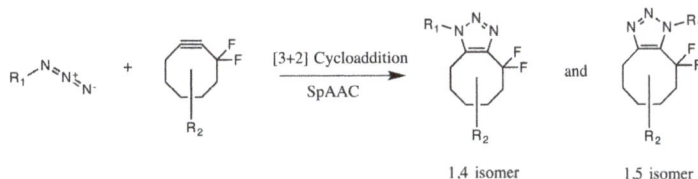

Scheme of the Strain-promoted Azide-Alkyne Cycloaddition

This reaction proceeds as a concerted [3+2] cycloaddition in the same mechanism as the Huisgen 1,3-dipolar cycloaddition. Substituents other than fluorines, such as benzene rings, are also allowed on the cyclooctyne.

This reaction has been used successfully to probe for azides in living systems, even though the reaction rate is somewhat slower than that of the CuAAC. Moreover, because the synthesis of cyclooctynes often gives low yield, probe development for this reaction has not been as rapid as for other reactions. But cyclooctyne derivatives such as DIFO, dibenzylcyclooctyne (DIBO) and biarylazacyclooctynone (BARAC) have all been used successfully in the SPAAC reaction to probe for azides in living systems.

Strain-promoted Alkyne-Nitrone Cycloaddition (SPANC)

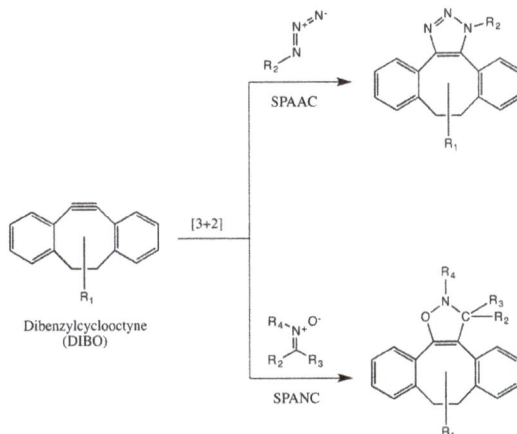

The SPAAC vs SpANC reaction

Diaryl-strained-cyclooctynes including dibenzylcyclooctyne (DIBO) have also been used to react with 1,3-nitrones in strain-promoted alkyne-nitrone cycloadditions (SPANC) to yield N-alkylated isoxazolines.

Because this reaction is metal-free and proceeds with fast kinetics (k2 as fast as 60 1/Ms, faster than both the CuAAC or the SPAAC) SPANC can be used for live cell labeling. Moreover, substitution on both the carbon and nitrogen atoms of the nitrone dipole, and acyclic and endocyclic nitrones are all tolerated. This large allowance provides a lot of flexibility for nitrone handle or probe incorporation.

However, the isoazoline product is not as stable as the triazole product of the CuAAC and the SpAAC, and can undergo rearrangements at biological conditions. Regardless, this reaction is still very useful as it has notably fast reaction kinetics.

The applications of this reaction include labeling proteins containing serine as the first residue: the serine is oxidized to aldehyde with NaIO$_4$ and then converted to nitrone with p-methoxybenzenethiol, N-methylhydroxylamine and p-ansidine, and finally incubated with Cyclooctyne to give a click product. The SPANC also allows for multiplex labeling.

Reactions of Strained Alkenes

Strained alkenes also utilize strain-relief as a driving force that allows for their participation in click reactions. Trans-cycloalkenes (usually cyclooctenes) and other strained alkenes such as oxanorbornadiene react in click reactions with a number of partners including Azides, Tetrazines and Tetrazoles. These reaction partners can interact specifically with the strained alkene, staying bioorthogonal to endogenous alkenes found in lipids, fatty acids, cofactors and other natural products.

Alkene and Azide [3+2] Cycloaddition

Oxanorbornadiene (or another activated alkene) reacts with azides, giving triazoles as a product. However, these product triazoles are not aromatic as they are in the CuAAC or SPAAC reactions, and as a result are not as stable. The activated double bond in oxanobornadiene makes a triazoline intermediate that subsequently spontaneously undergoes a retro Diels-alder reaction to release furan and give 1,2,3- or 1,4,5-triazoles. Even though this reaction is slow, it is useful because oxabornodiene is relatively simple to synthesize. The reaction is not, however, entirely chemoselective.

Alkene and Tetrazine Inverse-demand Diels-Alder

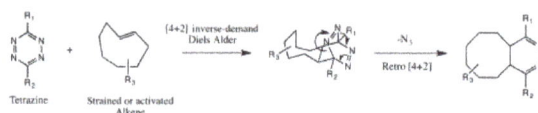

A Tetrazine-Alkene reaction between a generalized tetrazine and a strained, trans-cyclooctene

Strained cyclooctenes and other activated alkenes react with tetrazines in an inverse electron-demand Diels-Alder followed by a retro [4+2] cycloaddition. Like the other reactions of the trans-cyclooctene, ring strain release is a driving force for this reaction. Thus, three-membered and four-membered cycloalkenes, due to their high ring strain, make ideal alkene substrates.

Similar to other [4+2] cycloadditions, electron-donating substituents on the dienophile and electron-withdrawing substituents on the diene accelerate the inverse-demand diels-alder. The diene, the tetrazine, by virtue of having the additional nitrogens, is a good diene for this reaction. The dienophile, the activated alkene, can often be attached to electron-donating alkyl groups on target molecules, thus making the dienophile more suitable for the reaction.

Alkene and Tetrazole Photoclick Reaction

The Tetrazole-alkene "photoclick" reaction is another dipolar addition that Husigen first introduced about 50 years ago (ChemBioChem 2007, 8, 1504. (68) Clovis, J. S.; Eckell, A.; Huisgen, R.; Sustmann, R. Chem. Ber. 1967, 100, 60.) Tetrazoles with amino or styryl groups that can be activated by UV light at 365 nm (365 does not damage cells) react quickly (so that the UV light does not have to be on for a long time, usually around 1–4 minutes) to make fluorogenic pyrazoline products. This reaction scheme is well suited for the purpose of labeling in live cells, because UV light at 365 nm damages cells minimally. Moreover, the reaction proceeds quickly, so that the UV light can be administered for short durations. Finally, the non-fluorogenic reactants give rise to a fluorogenic product, equipping the reaction with a built-in spectrometry handle.

Both tetrazoles and the alkene groups have been incorporated as protein handles as unnatural amino acids, but this benefit is not unique. Instead, the photoinducibility of the reaction makes it a prime candidate for spatiotemporal specificity in living systems. Challenges include the presence of endogenous alkenes, though usually cis (as in fatty acids) they can still react with the activated tetrazole.

Applications

The applications of click chemistry are broad in that they allow for the attachment of a wide range of probes to a wide range of biomolecule targets. Several notable applications of click chemistry include the attachment of fluorescent probes for spectrometric quantification and qualification, and of handle molecules that allow for purification of the target biomolecule. For example, fluorophores such as rhodamine have been coupled onto norbonene, and reacted with tetrazine in living systems. In other cases, SPAAC between a cyclooctyne-modified fluorophore and azide-tagged proteins allowed the selection of these proteins in cell lysates.

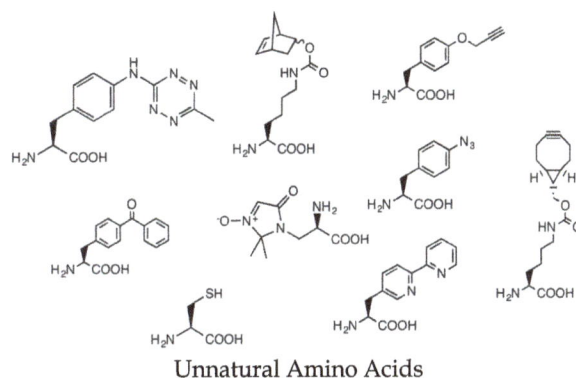
Unnatural Amino Acids

In addition, novel methods for the incorporation of click reaction partners into systems in and ex vivo contribute to the scope of possible reactions. The development of unnatural amino acid incorporation by ribosomes has allowed for the incorporation of click reaction partners as unnatural side groups on these unnatural amino acids. For example, an UAA with an azide side group provides convenient access for cycloalkynes to proteins tagged with this "AHA" unnatural amino acid. In another example, "CpK" has a side group including a cyclopropane alpha to an amide bond that serves as a reaction partner to tetrazine in an inverse diels-alder reaction.

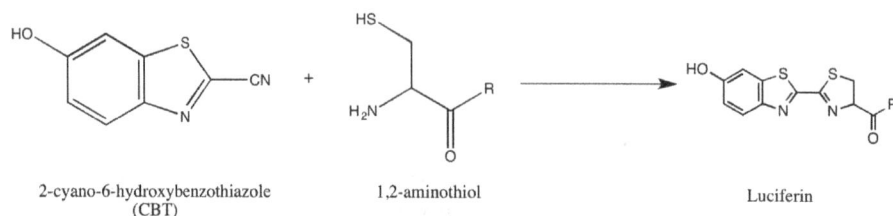

2-cyano-6-hydroxybenzothiazole   1,2-aminothiol   Luciferin
          (CBT)

Scheme of the synthesis of Luciferin

The synthesis of luciferin exemplifies another strategy of isolating reaction partners, which is to take advantage of rarely-occurring, natural groups such as the 1,2-aminothiol, which appears only when a cysteine is the final N' amino acid in a protein. Their natural selectivity and relative bio-orthogonality is thus valuable in developing probes specific for these tags. The above reaction occurs between a 1,2-aminothiol and a 2-cyanobenzothiazole to make luciferin, which is fluorescent. This luciferin fluorescence can be then quantified by spectrometry following a wash, and used to determine the relative presence of the molecule bearing the 1,2-aminothiol. If the quantification of non-1,2-aminothiol-bearing protein is desired, the protein of interest can be cleaved to yield a fragment with a N' Cys that is vulnerable to the 2-CBT.

Additional applications include:

- Two-dimensional gel electrophoresis separation

- preparative organic synthesis of 1,4-substituted triazoles

- modification of peptide function with triazoles

- modification of natural products and pharmaceuticals

- natural product discovery

- drug discovery

- macrocyclizations using Cu(I) catalyzed triazole couplings

- modification of DNA and nucleotides by triazole ligation

- supramolecular chemistry: calixarenes, rotaxanes, and catenanes

- dendrimer design

- carbohydrate clusters and carbohydrate conjugation by Cu(1) catalyzed triazole ligation reactions

- Polymers and Biopolymers

- surfaces

- material science

- nanotechnology, and

- Bioconjugation, for example, azidocoumarin.

- Biomaterials

In combination with combinatorial chemistry, high-throughput screening, and building chemical libraries, click chemistry has sped up new drug discoveries by making each reaction in a multistep synthesis fast, efficient, and predictable.

Technology License

The Scripps Research Institute has a portfolio of click-chemistry patents. Licensees include Invitrogen, Allozyne, Aileron, Integrated Diagnostics, and the biotech company baseclick, a BASF spin-off created to sell products made using click chemistry. Moreover, baseclick holds a worldwide exclusive license for the research and diagnostic market for the nucleic acid field. Fluorescent azides and alkynes also produced by such companies as Active Motif Chromeon and Cyandye

Biocompatibility

Biocompatibility is related to the behavior of biomaterials in various contexts. The term refers to the ability of a material to perform with an appropriate host response in a specific situation. The ambiguity of the term reflects the ongoing development of insights into how biomaterials interact with the human body and eventually how those interactions determine the clinical success of a medical device (such as pacemaker, hip replacement or stent). Modern medical devices and prostheses are often made of more than one material so it might not always be sufficient to talk about the biocompatibility of a specific material.

Indeed, since the immune response and repair functions in the body are so complicated it is not adequate to describe the biocompatibility of a single material in relation to a single cell type or tissue. Sometimes one hears of biocompatibility testing that is a large battery of in vitro test that is used in accordance with ISO 10993 (or other similar standards) to determine if a certain material (or rather biomedical product) is biocompatible. These tests do not determine the biocompatibility of a material, but they constitute an important step towards the animal testing and finally clinical trials that will determine the biocompatibility of the material in a given application, and thus medical devices such as implants or drug delivery devices.

IUPAC definition :

Biocompatibility (biomedical therapy): Ability of a material to perform with an appropriate host response in a specific application.

Biocompatibility: Ability to be in contact with a living system without producing an adverse effect.

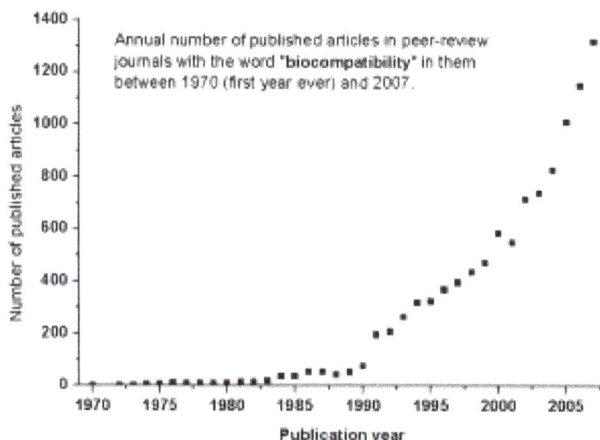

The word *biocompatibility* seems to have been mentioned for the first time in peer-review journals and meetings in 1970 by RJ Hegyeli (Amer Chem Soc Annual Meeting abstract) and CA Homsy et al. (J Macromol Sci Chem A4:3,615, 1970). It took almost two decades before it began to be commonly used in scientific literature.

Recently Williams (again) has been trying to reevaluate the current knowledge status regarding what factors determine clinical success. Doing so notes that an implant may not always have to be positively bioactive but it must not do any harm (either locally or systemically) (Williams, 2008).

Five definitions of Biocompatibility

1. "The quality of not having toxic or injurious effects on biological systems".

2. "The ability of a material to perform with an appropriate host response in a specific application", Williams' definition.

3. "Comparison of the tissue response produced through the close association of the implanted candidate material to its implant site within the host animal to that tissue response recognised and established as suitable with control materials" - ASTM

4. "Refers to the ability of a biomaterial to perform its desired function with respect to a medical therapy, without eliciting any undesirable local or systemic effects in the recipient or beneficiary of that therapy, but generating the most appropriate beneficial cellular or tissue response in that specific situation, and optimising the clinically relevant performance of that therapy".

5. "Biocompatibility is the capability of a prosthesis implanted in the body to exist in harmony with tissue without causing deleterious changes".

Comments on the above Five Definitions

1. The Dorland Medical definition not recommended according to Williams Dictionary since it only defines biocompatibility as the absence of host response and does not include any desired or positive interactions between the host tissue and the biomaterials.

2. This is also referred to as the "the Williams definition" or "William's definition". It was defined in the European Society for Biomaterials Consensus Conference I and can more easily be found in 'The Williams Dictionary of Biomaterials'.

3. The ASTM is not recommended according to Williams Dictionary since it only refers to local tissue responses, in animal models.

4. The fourth is an expansion or rather more precise version of the first definition noting both that low toxicity and the one should be aware of the different demands between various medical applications of the same material.

All these definitions deal with materials and not with devices. This is a drawback since many medical devices are made of more than one material. Much of the pre-clinical testing of the materials is not conducted on the devices but rather the material itself. But at some stage the testing will have to include the device since the shape, geometry and surface treatment etc. of the device will also affect its biocompatibility.

'Biocompatible'

In the literature, one quite often stumbles upon the adjective form, 'biocompatible'. However, according to Williams' definition, this does not make any sense because biocompatibility is contextual, i.e. much more than just the material itself will determine the clinical outcome of the medical device of which the biomaterial is a part. This also points to one of the weaknesses with the current definition because a medical device usually is made of more than one material.

Metallic glasses based on magnesium with zinc and calcium addition are tested as the potential biocompatible metallic biomaterials for biodegradable medical implants

Suggested Sub-definitions

The scope of the first definition is so wide that D Williams tried to find suitable subgroups of applications in order to be able to make more narrow definitions. In the MDT article from 2003 the chosen supgroups and their definitions were:

Biocompatibility of long-term implanted devices

> The biocompatibility of a long-term implantable medical device refers to the ability of the device to perform its intended function, with the desired degree of incorporation in the host, without eliciting any undesirable local or systemic effects in that host.

Biocompatibility of short-term implantable devices

> The biocompatibility of a medical device that is intentionally placed within the cardiovascular system for transient diagnostic or therapeutic purposes refers to the ability of the device to carry out its intended function within flowing blood, with minimal interaction between device and blood that adversely affects device performance, and without inducing uncontrolled activation of cellular or plasma protein cascades.

Biocompatibility of tissue-engineering products

The biocompatibility of a scaffold or matrix for a tissue-engineering products refers to the ability to perform as a substrate that will support the appropriate cellular activity, including the facilitation of molecular and mechanical signalling systems, in order to optimise tissue regeneration, without eliciting any undesirable effects in those cells, or inducing any undesirable local or systemic responses in the eventual host.

In these definitions the notion of biocompatibility is related to devices rather than to materials as compared to top three definitions. There was a consensus conference on biomaterial definitions in Sorrento September 15–16, 2005.

Bioconjugation

Bioconjugation is a chemical strategy to form a stable covalent link between two molecules, at least one of which is a biomolecule.

Function

Recent advances in the understanding of biomolecules enabled their application to numerous fields like medicine and materials. Synthetically modified biomolecules can have diverse functionalities, such as tracking cellular events, revealing enzyme function, determining protein biodistribution, imaging specific biomarkers, and delivering drugs to targeted cells. Bioconjugation is a crucial strategy that links these modified biomolecules with different substrates.

Synthesis

Synthesis of bioconjugates involves a variety of challenges, ranging from the simple and nonspecific use of a fluorescent dye marker to the complex design of antibody drug conjugates. As a result, various bioconjugation reactions – chemical reactions connecting two biomolecules together – have been developed to chemically modify proteins. Common types of bioconjugation reactions are coupling of lysine amino acid residues, coupling of cysteine residues, coupling of tyrosine residues, modification of tryptophan residues, and modification of the N- and C- terminus.

However, these reactions often lack chemoselectivity and efficiency, because they depend on the presence of native amino acid residues, which are usually present in large quantities that hinder selectivity. There is an increasing need for chemical strategies that can effectively attach synthetic molecules site specifically to proteins. One strategy is to first install a unique functional group onto a protein, and then a bioorthogonal or click type reaction is used to couple a biomolecule with this unique functional group. The bioorthogonal reactions targeting non-native functional groups are widely used in bioconjugation chemistry. Some important reactions are modification of ketone and aldehydes, Staudinger ligation with azides, copper-catalyzed huisgen cyclization of azide, strain promoted huisgen cyclization of azide.

Common Bioconjugation Reactions

The most common bioconjugations are coupling of a small molecule (such as biotin or a fluorescent dye) to a protein, or protein-protein conjugations, such as the coupling of an antibody to an enzyme. Other less common molecules used in bioconjugation are oligosaccharides, nucleic acids, synthetic polymers such as polyethylene glycol, and carbon nanotubes. Antibody-drug conjugates such as Brentuximab vedotin and Gemtuzumab ozogamicin are also examples of bioconjugation, and are an active area of research in the pharmaceutical industry. Recently, bioconjugation has also gained importance in nanotechnology applications such as bioconjugated quantum dots.

Reactions of Lysine Residues

The nucleophilic lysine residue is commonly targeted site in protein bioconjugation, typically through amine-reactive succinimidyl esters. To obtain optimal number of deprotonated lysine residues, the pH of the aqueous solution must be below the pKa of the lysine ammonium group, which is around 10.5, so the typical pH of the reaction is about 8 and 9. The common reagent for the coupling reaction is NHS-ester, which reacts with nucleophilic lysine through a lysine acylation mechanism. Other similar reagents are isocyanates and isothiocyanates that undergo a similar mechanism.

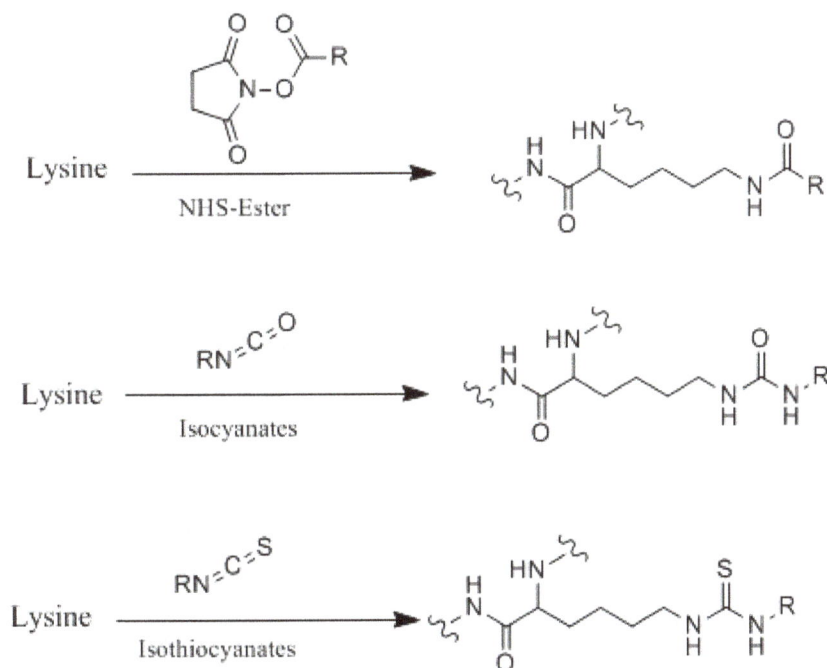

Figure 1. Bioconjugation strategies for lysine residues

Reactions of Cysteine Residues

Because free cysteine rarely occurs on protein surface, it is an excellent choice for chemoselective modification. Under basic condition, the cysteine residues will be deprotonated to generate a thiolate nucleophile, which will react with soft electrophiles, such as maleimides and iodoacetamides. As a result, a carbon-sulfur bond is formed. Another modification of cysteine residues involves the formation of disulfide bond. The reduced cysteine residues react with exogenous disulfides, generating new disulfides bond on protein. An excess of disulfides is often used to drive the reaction, such as 2-thiopyridone and 3-carboxy-4-nitrothiophenol.

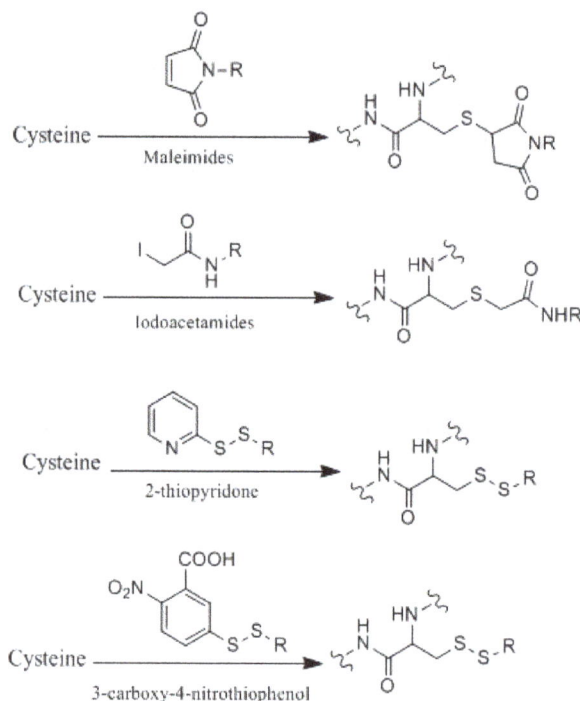

Figure 2. Bioconjugation strategies for cysteine residues

Reactions of Tyrosine Residues

Tyrosine residues are relatively unreactive; therefore they have not been a popular targets for bioconjugation. Recent development has shown that the tyrosine can be modified through electrophilic aromatic substitutions (EAS) reactions, and it is selective for the aromatic carbon adjacent to the phenolic hydroxyl group. This becomes particularly useful in the case that cysteine residues cannot be targeted. Specifically, diazonium effectively couples with tyrosine residues (diazonium salt shown as catalyst in Figure 3 below), and an electron withdrawing substituent in the 4-position of diazonium salt can effectively increase the efficiency of the reaction.

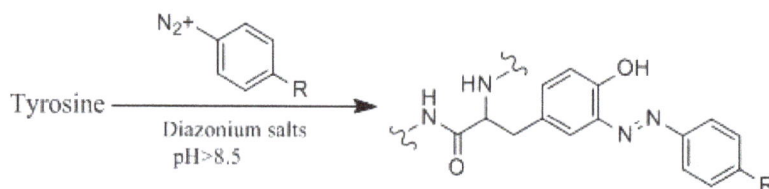

Figure 3. Bioconjugation strategies for tyrosine residues

Reactions of N- and C- termini

Since natural amino acid residues are usually present in large quantities, it is often difficult to modify one single site. Strategies targeting the termini of protein have been developed, because they greatly enhanced the site selectivity of protein modification. One of the N- termini modifications involves the functionalization of the terminal amino acid. The oxidation of N-terminal serine and threonine residues are able to generate N-terminal aldehyde, which can undergo further bioorthogonal reactions. Another type of modification involves the condensation of N-terminal cysteine with aldehyde, generating thiazolidine that is stable at high pH (second reaction in Figure 4). Using pyridoxal phosphate (PLP), several N-terminal amino acids can undergo transamination to yield N-terminal aldehyde, such as glycine and aspartic acid (third reaction in Figure 4).

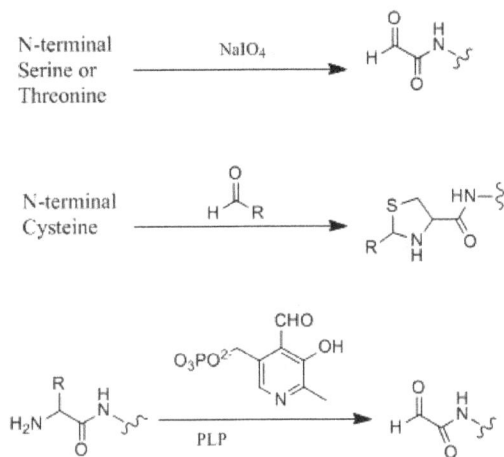

Figure 4. Bioconjugation strategies for N-terminus

An example of C-termini modification is the native chemical ligation (NCL), which is the coupling between a C-terminal thioester and a N-terminal cysteine (Figure 5).

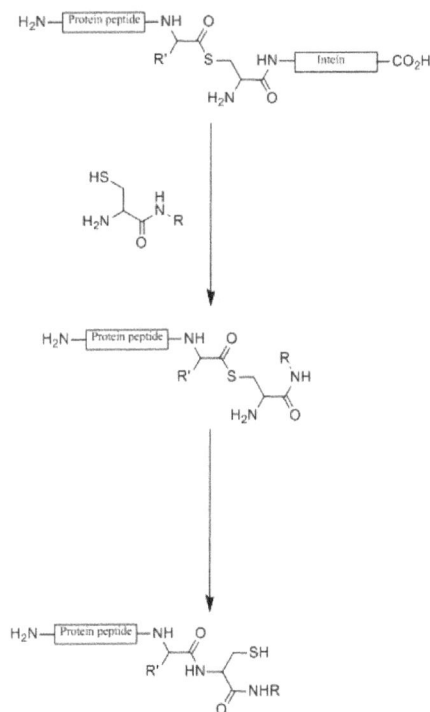

Figure 5. Bioconjugation strategies for C-terminus

Bioorthogonal Reactions

Modification of ketones and Aldehydes

A ketone or aldehyde can be attached to a protein through the oxidation of N-terminal serine residues or transamination with PLP. Additionally, they can be introduced by incorporating unnatural amino acids via the Tirrell method or Schultz method. They will then selectively condense with an alkoxyamine and a hydrazine, producing oxime and hydrazone derivatives. This reaction is highly chemoselective in terms of protein bioconjugation, but the reaction rate is slow. The mechanistic studies show that the rate determining step is the dehydration of tetrahedral intermediate, so a mild acidic solution is often employed to accelerate the dehydration step.

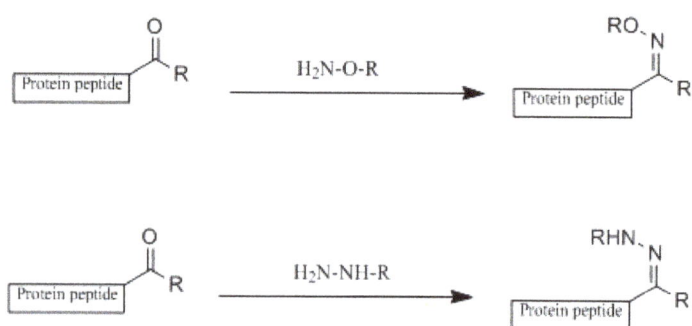

Figure 6. Bioconjugation strategies for targeting ketones and aldehydes

The introduction of nucleophilic catalyst can significantly enhance reaction rate. For example, using aniline as a nucleophilic catalyst, a less populated protonated carbonyl becomes a highly populated protonated Schiff base. In other words, it generates a high concentration of reactive electrophile. The oxime ligation can then occur readily, and it has been reported that the rate increased up to 400 times under mild acidic condition. The key of this catalyst is that it can generate a reactive electrophile without competing with desired product.

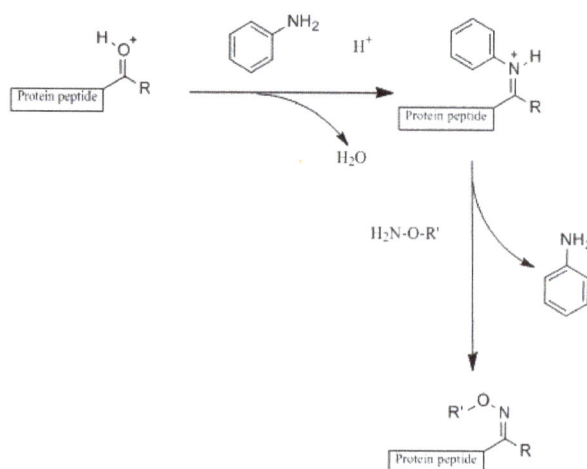

Figure 7. Nucleophilic catalysis of oxime ligation

Recent developments that exploit proximal functional groups have enabled hydrazone condensations to operate at 20 M^{-1}s^{-1} at neutral pH while oxime condensations have been discovered which proceed at 500-10000 M^{-1}s^{-1} at neutral pH without added catalysts.

Staudinger Ligation with Azides

The Staudinger ligation of azides and phosphine has been used extensively in field of chemical biology. Because it is able to form a stable amide bond in living cells and animals, it has been applied to modification of cell membrane, *in vivo*imaging, and other bioconjugation studies.

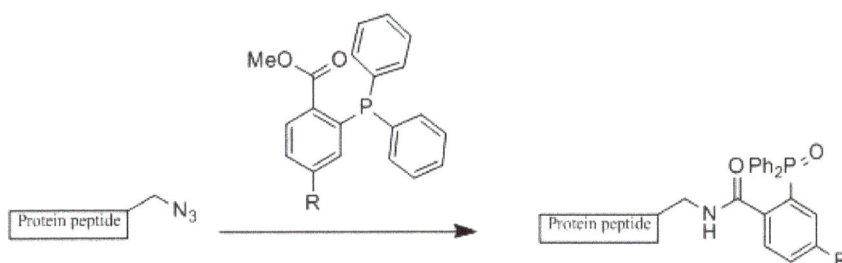

Figure 8. Staudinger Ligation with Azides

Contrasting with the classic Staudinger reaction, Staudinger ligation is a second order reaction in which the rate-limiting step is the formation of phosphazide (specific reaction mechanism shown in Figure 9). The triphenylphosphine first reacts with the azide to yield an azaylide through a four-membered ring transition state, and then an intramolecular reaction leads to the iminophosphorane intermediate, which will then give the amide-linkage under hydrolysis.

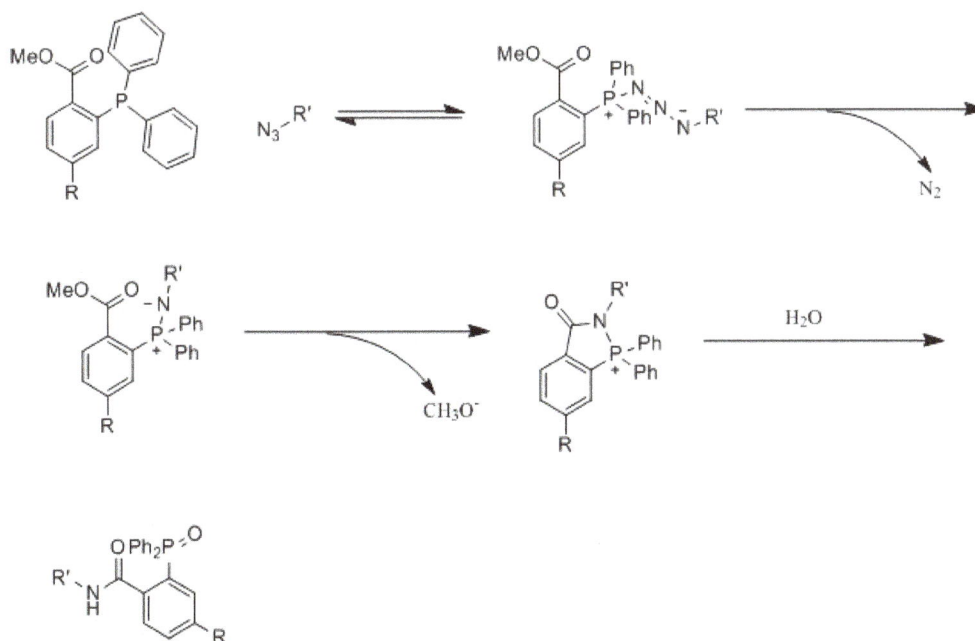

Figure 9.Mechanism of Staudinger Ligation

Copper Catalyzed Huisgen Cyclization of Azides

Azide has become a popular target for chemoselective protein modification, because they are small in size and has a favorable thermodynamic reaction potential. One such azide reactions is the [3+2] cycloaddition reaction with alkyne, but the reaction requires high temperature and often gives mixtures of regioisomers.

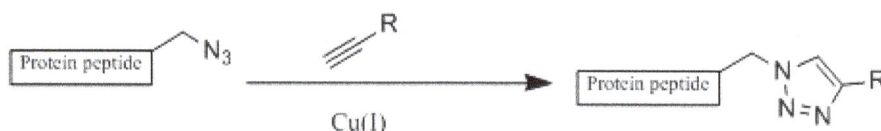

Figure 10. Copper-catalyzed cyclization of Azides

An improved reaction developed by chemist Karl Barry Sharpless involves the copper (I) catalyst, which couples azide with terminal alkyne that only give 1,4 substituted 1,2,3 triazoles in high yields (shown below in Figure 11). The mechanistic study suggests a stepwise reaction. The Cu (I) first couples with acetylenes, and then it reacts with azide to generate a six-membered intermediate. The process is very robust that it occurs at pH ranging from 4 to 12, and copper (II) sulfate is often used as a catalyst in the presence of a reducing agent.

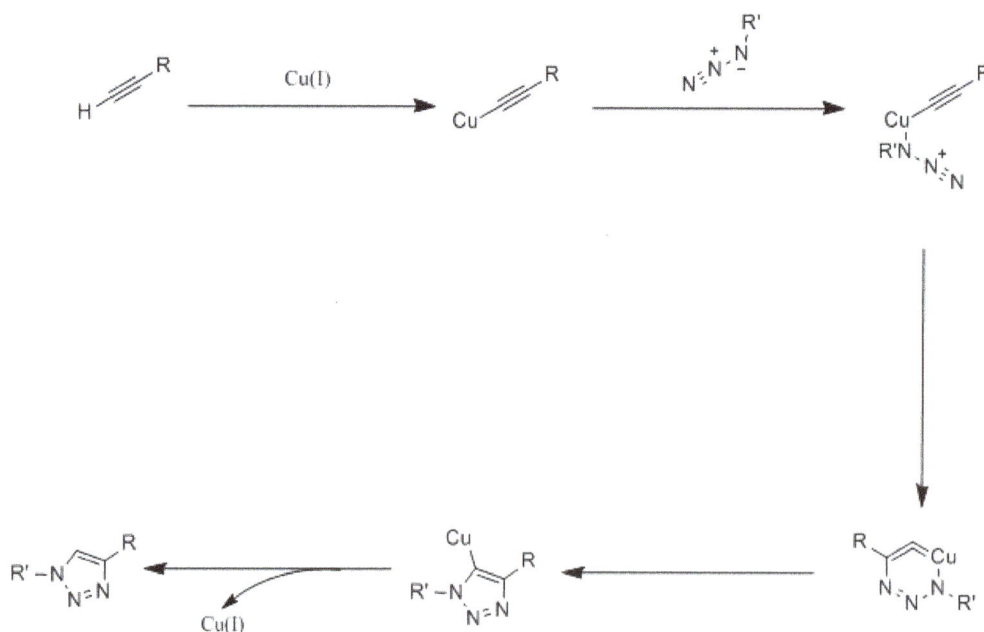

Figure 11. Mechanism for Copper-catalyzed cyclization of Azides

Strain Promoted Huisgen Cyclization of Azides

Even though Staudinger ligation is a suitable bioconjugation in living cells without major toxicity, the phosphine's sensitivity to air oxidation and its poor solubility in water significantly hinder its efficiency. The copper (I) catalyzed azide-alkyne coupling has reasonable reaction rate and efficiency under physiological conditions, but copper poses significant toxicity and sometimes inter-feres with protein functions in living cells. In 2004, chemist Carolyn R. Bertozzi's lab developed a metal free [3+2] cycloaddition using strained cyclooctyne and azide. Cyclooctyne, which is the smallest stable alkyne, can couple with azide through [3+2] cycloaddition, leading to two regioiso-meric triazoles (**Figure 12**). The reaction occurs readily in room temperature and therefore can be used to effectively modify living cells without negative effects. It has also been reported that the installation of fluorine substituents on cyclic alkyne can greatly accelerate the reaction rate.

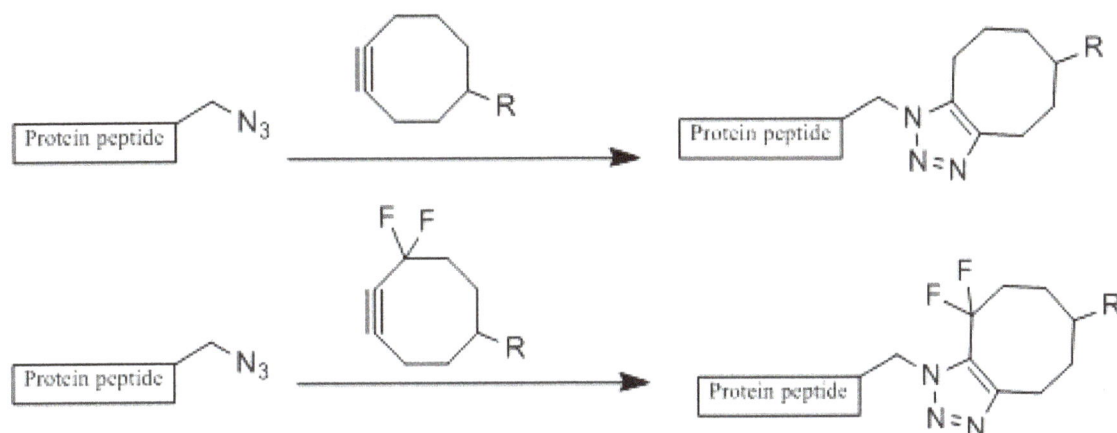

Figure 12. Strain promoted cycloaddition of azides and cyclooctynes

Azide-alkyne Huisgen Cycloaddition

The Azide-Alkyne Huisgen Cycloaddition is a 1,3-dipolar cycloaddition between an azide and a ter-minal or internal alkyne to give a 1,2,3-triazole. Rolf Huisgen was the first to understand the scope of this organic reaction. American chemist K. Barry Sharpless has referred to this cycloaddition as "the cream of the crop" of click chemistry and "the premier example of a click reaction."

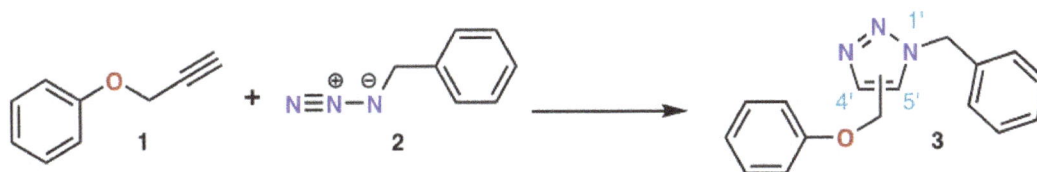

In the reaction above azide **2** reacts neatly with alkyne **1** to afford the triazole **3** as a mixture of 1,4-adduct and 1,5-adduct at 98 °C in 18 hours.

The standard 1,3-cycloaddition between an azide 1,3-dipole and an alkene as dipolarophile has largely been ignored due to lack of reactivity as a result of electron-poor olefins and elimination side reactions. Some success has been found with non-metal-catalyzed cycloadditions, such as the reactions using dipolarophiles that are electron-poor olefins or alkynes.

Although azides are not the most reactive 1,3-dipole available for reaction, they are preferred for their relative lack of side reactions and stability in typical synthetic conditions.

Copper Catalysis

A notable variant of the Huisgen 1,3-dipolar cycloaddition is the copper(I) catalyzed variant, no longer a true concerted cycloaddition, in which organic azides and terminal alkynes are united to afford 1,4-regioisomers of 1,2,3-triazoles as sole products (substitution at positions 1' and 4'). The copper(I)-catalyzed variant was first reported in 2002 in independent publications by Morten Meldal at the Carlsberg Laboratory in Denmark and Valery Fokin and K. Barry Sharpless at the Scripps Research Institute. While the copper(I)-catalyzed variant gives rise to a triazole from a terminal alkyne and an azide, formally it is not a 1,3-dipolar cycloaddition and thus should not be termed a Huisgen cycloaddition. This reaction is better termed the Copper(I)-catalyzed Azide-Alkyne Cycloaddition (CuAAC).

While the reaction can be performed using commercial sources of copper(I) such as cuprous bromide or iodide, the reaction works much better using a mixture of copper(II) (e.g. copper(II) sulfate) and a reducing agent (e.g. sodium ascorbate) to produce Cu(I) in situ. As Cu(I) is unstable in aqueous solvents, stabilizing ligands are effective for improving the reaction outcome, especially if tris-(benzyltriazolylmethyl)amine (TBTA) is used. The reaction can be run in a variety of solvents, and mixtures of water and a variety of (partially) miscible organic solvents including alcohols, DMSO, DMF, tBuOH and acetone. Owing to the powerful coordinating ability of nitriles towards Cu(I), it is best to avoid acetonitrile as the solvent. The starting reagents need not be completely soluble for the reaction to be successful. In many cases, the product can simply be filtered from the solution as the only purification step required.

NH-1,2,3-triazoles are also prepared from alkynes in a sequence called the Banert cascade.

The utility of the Cu(I)-catalyzed click reaction has also been demonstrated in the polymerization reaction of a bis-azide and a bis-alkyne with copper(I) and TBTA to a conjugated fluorene based polymer. The degree of polymerization easily exceeds 50. With a stopper molecule such as phenyl azide, well-defined phenyl end-groups are obtained.

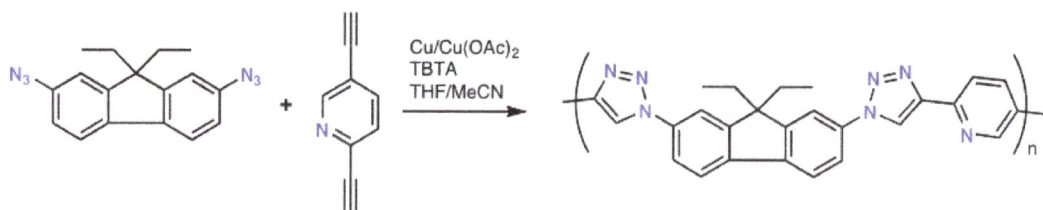

The copper-mediated azide-alkyne cycloaddition is receiving widespread use in material and surface sciences. Most variations in coupling polymers with other polymers or small molecules have been explored. Current shortcomings are that the terminal alkyne appears to participate in free radical polymerizations. This requires protection of the terminal alkyne with a trimethyl silyl protecting group and subsequent deprotection after the radical reaction are completed. Similarly the use of organic solvents, copper (I) and inert atmospheres to do the cycloaddition with many polymers makes the "click" label inappropriate for such reactions. An aqueous protocol for performing the cycloaddition with free radical polymers is highly desirable.

The CuAAC click reaction also effectively couples polystyrene and bovine serum albumin (BSA). The result is an amphiphilic biohybrid. BSA contains a thiol group at Cys-34 which is functionalized with an alkyne group. In water the biohybrid micelles with a diameter of 30 to 70 nanometer form aggregates.

Copper Catalysts

The use of a Cu catalyst in water was an improvement over the same reaction first popularized by Rolf Huisgen in the 1970s, which he ran at elevated temperatures. The traditional reaction is slow and thus requires high temperatures. However, the azides and alkynes are both kinetically stable.

As mentioned above, copper-catalysed click reactions work essentially on terminal alkynes. The Cu species undergo metal insertion reaction into the terminal alkynes. The Cu(I) species may either be introduced as preformed complexes, or are otherwise generated in the reaction pot itself by one of the following ways:

- A Cu compound (in which copper is present in the +2 oxidation state) is added to the reaction in presence of a reducing agent (e.g. sodium ascorbate) which reduces the Cu from the (+2) to the (+1) oxidation state. The advantage of generating the Cu(I) species in this manner is it eliminates the need of a base in the reaction. Also the presence of reducing agent makes up for any oxygen which may have gotten into the system. Oxygen oxidises the Cu(I) to Cu(II) which impedes the reaction and results in low yields. One of the more commonly used Cu compounds is $CuSO_4$

- Oxidation of Cu(0) metal

- Halides of copper may be used where solubility is an issue. However, the iodide and bromide Cu salts require either the presence of amines or higher temperatures.

Commonly used solvents are polar aprotic solvents such as THF, DMSO, Acetonitrile, DMF as well as in non-polar aprotic solvents such as toluene. Neat solvents or a mixture of solvents may be used.

DIPEA (N,N-Diisopropylethylamine) and Et_3N (triethylamine) are commonly used bases.

Mechanism

A mechanism for the reaction has been suggested based on density functional theory calculations. Copper is a 1st row transition metal. It has the electronic configuration [Ar] $3d^{10}$ $4s^1$. The copper (I) species generated in situ forms a pi complex with the triple bond of a terminal alkyne. In the presence of a base, the terminal hydrogen, being the most acidic is deprotonated first to give a Cu acetylide intermediate. Studies have shown that the reaction is second order with respect to Cu. It has been suggested that the transition state involves two copper atoms. One copper atom is bonded to the acetylide while the other Cu atom serves to activate the azide. The metal center coordinates with the electrons on the nitrogen atom. The azide and the acetylide are not coordinated to the same Cu atom in this case. The ligands employed are labile and are weakly coordinating. The azide displaces one ligand to generate a copper-azide-acetylide complex. At this point cyclisation takes place. This is followed by protonation; the source of proton being the hydrogen which was pulled off from the terminal acetylene by the base. The product is formed by dissociation and the catalyst ligand complex is regenerated for further reaction cycles.

The reaction is assisted by the copper, which, when coordinated with the acetylide lowers the pKa of the alkyne C-H by up to 9.8 units. Thus under certain conditions, the reaction may be carried out even in the absence of a base.

In the uncatalysed reaction the alkyne remains a poor electrophile. Thus high energy barriers lead to slow reaction rates.

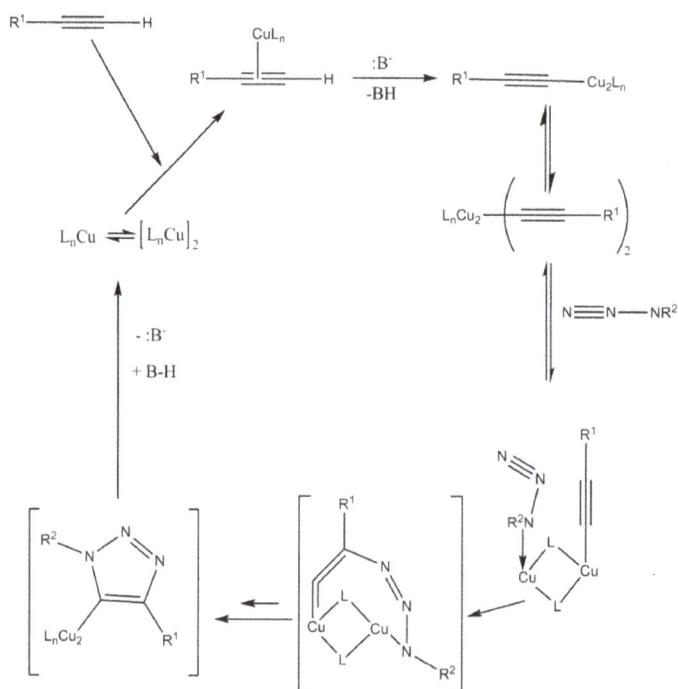

Figure 1: Copper catalysed Click reaction

Ligand Assistance

The ligands employed are usually labile i.e. they can be displaced easily. Though the ligand plays no direct role in the reaction the presence of a ligand has its advantages. The ligand protects the Cu ion from interactions leading to degradation and formation of side products and also prevents the oxidation of the Cu(I) species to the Cu(II). Furthermore, the ligand functions as a proton acceptor thus eliminating the need of a base.

Ruthenium Catalysis

The ruthenium-catalysed 1,3-dipolar azide-alkyne cycloaddition (RuAAC) gives the 1,5-triazole. Unlike CuAAC in which only terminal alkynes reacted, in RuAAC both terminal and internal alkynes can participate in the reaction. This suggests that ruthenium acetylides are not involved in the catalytic cycle.

The proposed mechanism suggests that in the first step, the spectator ligands undergo displacement reaction to produce an activated complex which is converted, via oxidative coupling of an alkyne and an azide to the ruthenium containing metallocyle (Ruthenacycle). The new C-N bond is formed between the more electronegative and less sterically demanding carbon of the alkyne and the terminal nitrogen of the azide. The metallacycle intermediate then undergoes reductive elimination releasing the aromatic triazole product and regenerating the catalyst or the activated complex for further reaction cycles.

Cp*RuCl(PPh$_3$)$_2$, Cp*Ru(COD)and Cp*[RuCl$_4$] are commonly used ruthenium catalysts. Catalysts containing cyclopentadienyl(Cp) group are also used. However, better results are observed with the pentamethylcyclopentadienyl(Cp*) version. This may be due to the sterically demanding Cp* group which facilitates the displacement of the spectator ligands.

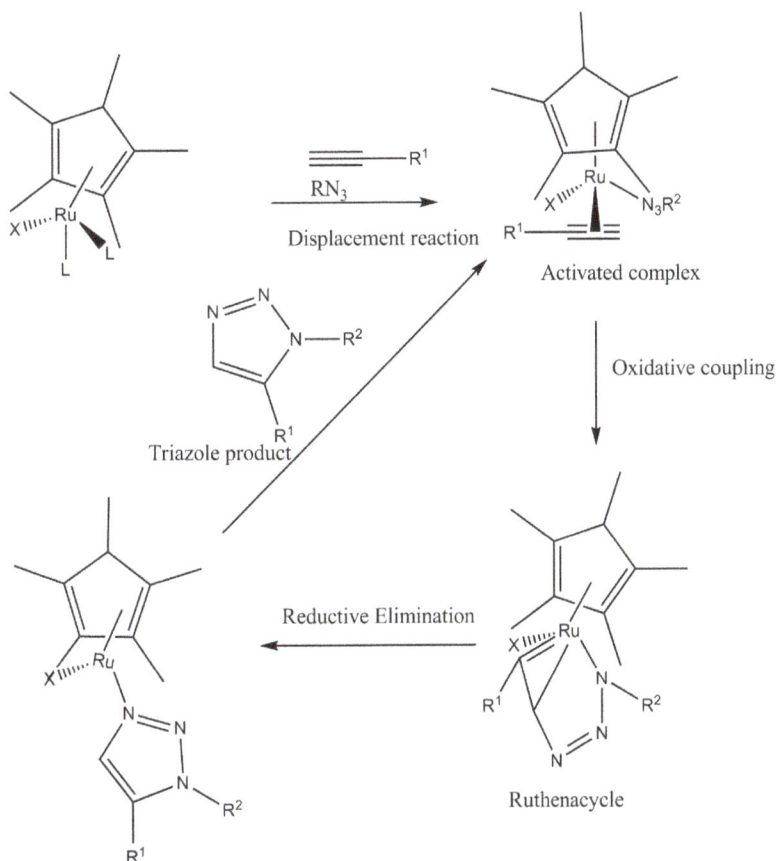

X = Cl
L = Spectator ligand

Figure 2: Mechanism of Ruthenium catalysed click reaction

Silver Catalysis

Recently, the discovery of a general Ag(I)-catalyzed azide–alkyne cycloaddition reaction (Ag-AAC) leading to 1,4-triazoles is reported. Mechanistic features are similar to the generally accepted mechanism of the copper(I)-catalyzed process. Interestingly, silver(I)-salts alone are not sufficient to promote the cycloaddition. However the ligated Ag(I) source has proven to be exceptional for AgAAC reaction. Curiously, pre-formed silver acetylides do not react with azides; however, silver acetylides do react with azides under catalysis with copper(I).

References

- Tilley, S. D.; Joshi, N. S.; Francis, M. B. (2008). "Proteins: Chemistry and Chemical Reactivity". Wiley Encyclopedia of Chemical Biology. doi:10.1002/9780470048672.wecb493. ISBN 0470048670.

- "Biocompatibility Safety Assessment of Medical Devices: FDA/ISO and Japanese Guidelines". Mddionline.com. Retrieved 20 November 2014.

- Douglas A MacKenzie1,2, Allison R Sherratt1, Mariya Chigrinova1,2, Lawrence LW Cheung1,2 and John Paul Pezacki1,2, 2014, 21:81-88.

- "Terminology for biorelated polymers and applications (IUPAC Recommendations 2012)" (PDF). Pure and Applied Chemistry. 84 (2): 377–410. 2012. doi:10.1351/PAC-REC-10-12-04.

- L. Liang and D. Astruc: "The copper(I)-catalysed alkyne-azide cycloaddition (CuAAC) "click" reaction and its applications. An overview", 2011, 255 , 23-24, 2933-2045, p. 2934

- Francis, M. B.; Carrico, I. S. (2010). "New frontiers in protein bioconjugation". Current Opinion in Chemical Biology. 14 (6): 771–773. doi:10.1016/j.cbpa.2010.11.006. PMID 21112236.

- Kalia, J.; Raines, R. T. (2010). "Advances in Bioconjugation". Current organic chemistry. 14 (2): 138–147. PMC 2901115 . PMID 20622973.

- Saxon, E.; Bertozzi, C. R. (2000). "Cell Surface Engineering by a Modified Staudinger Reaction". Science. 287 (5460): 2007–2010. doi:10.1126/science.287.5460.2007. PMID 10720325.

Permissions

Index

www.ingramcontent.com/pod-product-compliance
Lightning Source LLC
Chambersburg PA
CBHW082058190326
41458CB00010B/3528